Ships' Bilge Pumps

NUMBER TWO:
Studies in Nautical Archaeology

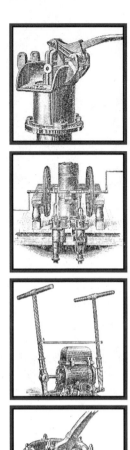

Ships' Bilge Pumps

A History of Their Development, 1500–1900

THOMAS J. OERTLING

TEXAS A&M UNIVERSITY PRESS
COLLEGE STATION

Copyright © 1996 by Thomas J. Oertling
Manufactured in the United States of America
All rights reserved
First edition

The paper used in this book meets the minimum requirements
of the American National Standard for Permanence
of Paper for Printed Library Materials, Z39.48-1984.
Binding materials have been chosen for durability.

Library of Congress Cataloging-in-Publication Data
Oertling, Thomas J. (Thomas James), 1954–
    Ships' bilge pumps : a history of their development, 1500–1900 /
Thomas J. Oertling.— 1st ed.
       p.   cm. — (Studies in nautical archaeology ; no. 2)
    Includes bibliographical references (p. –) and index.
    ISBN 0-89096-722-9
    1. Marine pumps—History. I. Title. II. Series.
VM821.O37   1996
623.8'73—dc20   96-21939
                                                                   CIP

A.M.D.G.

To my parents
Sewall and Lillian Oertling
who told me to do what would make me happy

Contents

	PAGE
List of Illustrations	IX
Acknowledgments	XIII
Introduction	XV

CHAPTERS
1.	Of Leaks and Men	3
2.	The Construction of Wooden Tubes	10
3.	The Burr Pump	16
4.	Common or "Suction" Pumps	22
5.	The Chain Pump	56
6.	Later Pump Types	74
7.	Summary and Conclusions	79

Notes	83
Bibliography	91
Index	99

Illustrations

	Figures	PAGE
1.	Boring a wooden tube.	11
2.	Shell augers and bits.	12
3.	The parts of a burr pump.	17
4.	The foot valve and leather valve claque from the Basque whaler *San Juan*.	18
5.	The pump spear and leather valve from the *San Juan*.	18
6.	A typical sixteenth-century pump seating.	19
7.	The burr valve and spear from the mid-nineteenth-century wreck of a canal sloop found near Isle La Motte, Lake Champlain.	20
8.	Burr valve and spear from the wreck of the canal sloop near Isle La Motte, Lake Champlain.	20
9.	Schematic representation of a common pump.	24
10.	Schematic representation of a common pump's lower-valve body.	25
11.	The body of a lower piston valve from the *Machault*.	26
12.	Renderings of two types of upper valves.	27
13.	Two upper-valve bodies, a leather gasket, and two lower-valve bodies from the *Machault*.	28
14.	An upper-valve body from the *Machault*.	28

15. A typical common pump seated between the frames. 31
16. The lower end of one of the asymmetrical pump tubes from *El Nuevo Constante*. 32
17. Common-pump sieve from *El Nuevo Constante*. 33
18. One of the two lead disks recovered from the Molasses Reef wreck. 34
19. Common pump upper valve from the Molasses Reef wreck. 35
20. A woodcut from Agricola's *De Re Metallica*. 36
21. Detail of fig. 20. 37
22. The lead pump from the *San José*, resting upside down. 38
23. The *San José*'s pump reservoir and piston tube. 38
24. The lower extremity of the *San José*'s piston tube. 38
25. Lead pump components from the York River. 39
26. A lead and copper pump (RS 6) recovered near Saint Thomas. 40
27. Copper pump tube with lead nozzle from the American China trader *Rapid*. 41
28. *Corps de fonte* of a French royal pump recovered from the Fort Louisbourg wreck. 42
29. The bronze upper-piston valve of a French royal pump from the Fort Louisbourg wreck. 43
30. Details of the bronze upper-piston valve from the Fort Louisbourg wreck. 43
31. Dodgeson's pump of 1799. 46
32. Taylor's pump (1780). 47
33. One way a double-piston pump could be operated. 48
34. A double-piston valve from the wreck of HMS *Tribune*. 48
35. Details of the double-piston valve from the wreck of the *Tribune*. 49
36. Elements of the double-piston pump from the wreck of HMS *De Braak*. 50

37.	Schematic of the double-piston valve assembly from HMS *De Braak*'s pump.	51
38.	Typical arrangement of pumps on large sailing vessels from the mid-nineteenth century into the twentieth.	52
39.	An Admiralty plan dated 1807 for a seventy-four-gun ship.	53
40.	An 1825 plan of a wash pump set into a cistern mounted on brackets in the well.	54
41.	The Cole-Bentinck chain pump.	57
42.	Sprocket and drive wheel of an early type of chain pump.	62
43.	Chain and valve assembly of an early type of chain pump.	64
44.	Cutaway view of the *Santo António de Tanna*'s reconstructed chain pump.	65
45.	View from the port quarter of the midships area of the *Charon*.	66
46.	The seating of HMS *Charon*'s chain pump.	67
47.	Chain assembly of HMS *Charon*'s chain pump.	68
48.	The valve assembly of HMS *Charon*'s chain pump.	71
49.	Representation of the chain and valve assembly on the *Victory* and the *Pomone*.	72
50.	New Deluge bilge pump.	75
51.	Two-cylinder brass force pump.	76
52.	Two-cylinder brass force pump at the Penobscot Marine Museum.	76
53.	Challenge and Alert double-acting force pumps.	76
54.	Parts for the Edson diaphragm pump.	77
55.	Edson diaphragm pump.	77

Table

1.	Dimensions of French *Corps des Fontes*.	44

Acknowledgments

It was only with the help of a great many people that I was able to collect the many bits and pieces of information this work comprises. To those who supplied me with material or extended their help, both for my original thesis work and for this new edition, I extend my sincere thanks.

I would like to acknowledge the assistance and interest of the following people and institutions: Robert Grenier, Peter Waddell, and Walter Zacharchuk of the Canadian government's Underwater Archaeology Section of Environment/Parks; Steven Brooke of the Maine State Museum; Irving Kallock of the Saco Pump Museum; Paul Johnston and the Peabody Museum of Salem; James Blackaby and the Mercer Museum; John Sands and the Mariners' Museum; Charles Pearson and Coastal Environments, Inc.; Chuck Fithian, Delaware State Museums; John Broadwater and the Virginia Research Center for Archaeology; Robin Piercy, Donald Keith, and the Institute of Nautical Archaeology at Texas A&M University. I would also like to thank the National Maritime Museum in Greenwich, England, for the use of materials from its collections. J. J. Berrier, John Bingeman, Graeme Henderson, Brian Lavery, James Levy, David Lyon, John P. Oleson, and Alan Roddie also provided me with information.

I wish to thank the selection committee of the Studies in Nautical Archaeology Series for choosing my thesis to be a part of this series.

For their kind hospitality, I am grateful to Chris and Marzena Amer, Carl and Penny Hector, Margaret and Mr. and Mrs. Wilson

Morden, Mark and Debbie Palmer, Harding Polk and Roni Hinote, Warren Riess and Sheli Smith, Dory Slane, Dick and Sharon Swete, and Edward and Fay Tiencken.

Very special thanks go to my dear friends and colleagues, Roger and K. C. Smith, Mike Fitzgerald, Kevin Crisman, Bill Bayreuther, Joe J. Simmons, Denise Lakey, and Cheri Haldane, for all their help. The original thesis would never have been finished without the support and love of my parents and family. Of course, for this present work, I wish to thank my wife, Sarita, for her love and support during this second go-round.

Introduction

One of the more important pieces of equipment required for the effective and efficient operation of a ship is the bilge pump. All ships leak, so some means of expelling excess water from within the hull is needed. In small open boats a bailing bucket or scoop is sufficient, but when a vessel has a deep draft or if a deck encloses the hold, it becomes difficult—even impossible—to raise the water over the bulwarks by hand. Under such circumstances, a pump is a virtual necessity.

Although there is a large corpus of information pertaining to the guns, sails, rigging, and construction of ships that sailed between about 1500 and the mid-nineteenth century, little attention has been devoted to their bilge pumps. Most modern works regarding ships and seafaring make only passing references to pumps and do not go much beyond a general description of the different types. Yet these mundane machines were more important than the anchors, rudder, or sails, because they preserved the hull's buoyancy—without which a ship becomes a shipwreck. As an example, the Dutch merchantman *Vrouw Maria*'s (1771)[1] anchor held, but the ship sank because the pumps failed to keep up with the rising water.[2]

The scope of this work encompasses the use of the burr pump, the common or "suction" pump, and the chain pump on ships from ca. 1500 to ca. 1840. The force and chain pumps of Roman and early Byzantine date are becoming better understood, but otherwise there is a serious dearth of information on pumps from earlier times. Historical information on pumping machinery is also scarce. Thomas

Ewbank's *A Descriptive Historical Account of Hydraulic and Other Machines for Raising Water* (1842)[3] is a comprehensive text that addresses all types of hydraulic machines. Ewbank had access to a wide selection of eighteenth- and nineteenth-century texts, and he documented most of his references. In addition, he included his own observations regarding a number of ships' pumps. For these reasons, his text has been a primary source of information and references. Abbot Usher, in his discussion of "Machines and Mechanisms" (1957), uses pumps to illustrate the development of technology in the sixteenth and seventeenth centuries. He points out the increasing use of metals (lead, copper, and iron) during this time but notes that the technology of metalworking proceeded slowly and that cost was an important factor.

Contemporary marine dictionaries are helpful in supplying definitions, descriptions, and dates of use. When editions from different years are available, they can be compared to show changes in pump use through time (Blanckley 1732, 1755; Boteler 1634; Falconer 1776, 1780, 1815; Manwayring 1644; Smith 1627). Other contemporary texts provide accounts of routine use and additional information about designs and materials (Agricola 1556; Belidor 1737–53; Diderot 1966; Hutchinson 1794; *Treatise* 1793). Of course other books, monographs, and museum collections have been consulted (Boudriot 1974; Bugler 1966; Cederlund 1981, 1982; Longridge 1955; Edlin 1949; Rose 1937; Prager and Scaglia 1972). Only a few useful pieces of archival and patent information could be obtained from American and English sources.[4] A *brief* survey of material available in the U.S. National Archives yielded only one description of a pump and several shipyard account ledgers of repairs, which mention the purchase of pump leather and pump nails. Copies of a number of drawings of ships' pumps were purchased from the National Maritime Museum in Greenwich. They illustrate chain pumps, suction pumps of various designs, and arrangements of wash pumps. Additional information is found in the Admiralty Collections, Letters from the Navy Board, in the National Maritime Museum. Robert Gardiner (1976) based an article on these and other letters.

Finally, shipwreck accounts are useful sources because they give some idea of how pumps were operated in cases of extreme emergency (Duffy 1955; Huntress 1974).

The original study ended at 1840 because by then the manufacture of pumps was becoming significantly industrialized; with the advent of commercial steamships, more and more pumps were being driven by engines and undergoing radical technological advances. Furthermore, new designs were recorded in patent records, advertisements, and catalogs of the late nineteenth century. Such mass-produced and well-documented machines were therefore outside the technological and archaeological scope of the original investigation. Chapter 6 has been included to discuss some pump types that appeared in the late nineteenth and early twentieth centuries.

Ships' Bilge Pumps

1

Of Leaks and Men

On the King's birthday we put to sea.
How I wish I was in Sherbrook now.
We were ninety-one days to Montego Bay,
Pumping like madmen all the way,
God damn them all.

Stan Rogers,
"Barret's Privateers"

This work began as a survey of artifacts. While researching accounts of sinkings and other maritime disasters for descriptions of pumps, however, I came to appreciate the experiences, fears, ingenuity, intelligence, and resourcefulness of mankind represented by these artifacts. I realized that this study would not be complete without—and in fact draws much of its importance from—an understanding of the human responses to the escalating terror of a sinking ship.

Before investigating these responses, it would be beneficial to examine the nature of leaks in ships and the methods used to locate and control them. The eighteenth-century naval architect William Hutchinson[1] observed that sometimes the crews of merchantmen left their sinking ships too soon. Although these ships would appear to be in imminent danger of foundering, they were often discovered hours later, still afloat. There are two explanations of Hutchinson's observation. First, as water rises in the hold, the distance it must be lifted for removal is decreased, making it easier for tired crews to work the pumps and keep pace with the leaks. Second, after the water in the hold rises enough to cover the leak or leaks, the rate of the influx is reduced.[2] The proportional rate of water flow varies as the square root of the depth of the hole below the waterline. That is, a hole sixteen feet below the waterline admits water four times faster than a hole one foot below the waterline, given holes of equal size and similar position in the hull. But after the water within the hull covers the hole, the proportional rate of flow varies as the square root of the distance between the water within the hull and the waterline. At some point, therefore, a state of equilibrium can be reached between the volume of water being pumped or bailed out and the volume of water entering the hull.[3] For this reason it was possible to keep the ship afloat through continual bailing and pumping, even with a great amount of water in the hold. Some accounts of ship disasters relate that serious conditions were thus held in check until the ship could reach shore, provided the pump or pumps remained in working order.[4]

The most common type of leak arose along a planking seam whose oakum (the hemp or frayed rope used to caulk seams) had worked loose as a result of the torsional stresses exerted on the ship as it moved through the water. Such leaks could be found simply by listening for them, often by using some object to amplify the sound transmitted through the planking. By placing the mouth of an empty earthen pot against the planking and putting one's ear to the pot's bottom, the leak could be heard as a low rumbling. A short stick or even a trumpet could be used in the same way. The leak was located by moving the listening device to where the noise was loudest.[5]

After the leak had been found, it could be plugged or stopped in various ways, depending on its size, severity, position in the hull, the materials at hand, and the degree of panic among the crew and passengers. Ideally, the captain would prefer to put his ship in dry dock or to careen her in order to repair the bottom. But leaks had a nasty way of appearing in the course of a voyage, often during a storm. Stopgap methods were employed to eliminate or control them until a safe port could be reached.

Leaks could be plugged from inside the hull, especially in inaccessible areas, by "sinking down some tallow and coals mixed together." When the leak was very large, "pieces of raw beef, oatmeal bags, and the like stuff" were used to fill the hole.[6] During a voyage to the Orient, the crew of the Portuguese vessel *Saõ Thomé* (1589) plugged a severe leak with sacks of rice and placed a chest on top of them to weight the patch against the force of the water.[7] Apparently, salted beef was an adequate caulking medium, since it was used thus on an early voyage to the American colonies.[8] Indeed, after several weeks or months in the cask, the properties of salted beef were probably better suited to filling planking seams than men's stomachs.

For leaks between wind and water (i.e., in the vicinity of the waterline), a piece of sheet lead could be nailed over the problem area.[9] Later on, two layers of canvas or leather, backed with oakum, were recommended instead of lead because the latter tended to fatigue and crack as the hull worked.[10] Even so, lead strips were found on wrecks of the 1554 fleet at Padre Island, Texas, and were in place on the planking seams of one of the hulls. The strips had been tacked on top of resin-soaked cloth that overlay caulking consisting of hair and oakum, presumably to seal a leaking seam.[11] Shot holes at or just below the waterline were usually stopped with a canvas-covered wooden plug that was driven into the hole.[12]

Leaks were plugged from the outside of the hull by covering the area with a patch of sail sewn with oakum or by lowering a bag or net of oakum down on a line until it passed over the leak. The force of the water flowing through the hole into the ship pushed strands of the oakum into the breach, thereby arresting leakage. This worked best if the leak was in the bow or midships area.[13]

Not only does a ship float, but many of its contents do too. The account of the wreck of HMS *Centaur* (1782) reveals the havoc floating debris can wreak. This seventy-four-gun third-rate was returning to England from the West Indies leaking badly after the Battle of the Saints in 1782. Struck by a storm in the Atlantic, she was dismasted and lost her rudder. The ship's carpenter, however, addressed neither problem but devoted himself to keeping the pumps working. The ship's fate was sealed when debris smashed the well and displaced the pumps. Archaeologists should note well the kinetics of a sinking ship; this account bears important implications for the identification of in situ or secondary deposits of materials, the interpretation of interior structures, and the nature of the sinking and breakup of the hull.

> All the rum, twenty-six puncheons, and all the provisions of which there was sufficient for two months, in casks, were staved having floated with violence from side to side until there was not a whole cask remaining: even the staves that were found upon clearing the

hold were most of them broken in two pieces. In the fore-hold, we had a prospect of perishing: should the ship swim, we had no water but what remained in the ground tier; and over this all the wet provisions, and butts filled with salt water, were floating, and with so much motion that no man could with safety go into the hold. . . . What I have called the wreck of the hold, was the bulkheads of the after-hold, fishroom, and spirit rooms. The standards of the cockpit, an immense quantity of staves and wood and part of the lining of the ship were thrown overboard, that if the water should again appear in the hold, we might have no impediment in bailing. . . . At this period the carpenter acquainted me the well was staved in, destroyed by the wreck of the hold, and the chain pumps displaced and totally useless. There was nothing left but to redouble our efforts in bailing, but it became difficult to fill the buckets from the quantity of staves, planks, anchor-stock, and yardarm pieces, which now washed from the wings, and floating from side to side with the motion of the ship.[14]

Although this study focuses upon the details of pump design, periods of use, and developments in technology and material usage, one should not lose touch with the human factor. Living conditions aboard a sailing ship were poor, sanitation worse.[15] The inevitable accumulation of filth and garbage in the hold polluted the bilges, and although a health hazard, the nature of the bilge water did provide ready proof of whether the hull was "tight."

When a ship is staunch, that is takes in but little water into her hold, she is said to be tight. And this tightness is best known by the very smell of the water that is pumped out of her; for when it stinketh much, it is a sign that the water hath lain long in the hold of the ship; and on the contrary, when it is clear and sweet, it is a token that it comes freshly in from the sea. This stinking water therefore is always a welcome perfume to an old seaman; and he that stops his nose at it is laughed at, and held but a fresh-water man at best.[16]

The seamen certainly didn't enjoy the smell of the bilge but simply accepted it as a part of their life:

For when that we shall go to bedde,
The pump was nygh our beddes hede;

> *A man were as good as dede*
> *As smell thereof the stynk.*[17]

Even the bad smells were hazards. García de Palacio warned that before anyone was to enter the pump well (the compartment or room enclosing the pump), a lantern with a burning candle should be lowered down into it. If the candle went out, it was a sign of bad air and meant anyone entering the well might die, "as has happened."[18]

The possibility of the ship sinking must be acknowledged by every seafarer at some point (when the last mooring line is cast off, for instance). What does one think of upon hearing the phrase "to have a sinking feeling"? On land it may suggest acute anxiety or an uncomfortable feeling in the pit of one's stomach. At sea it could more vividly mean the abandonment of all hope, to become resigned to one's fate upon a merciless sea. The words of an eyewitness best convey the feelings of those who have experienced such dire straits:

> All this night they passed in great trouble and distress, for everything they could see represented death. For beneath them they saw a ship full of water, and above them the Heavens conspired against all, for the sky was shrouded with the deepest gloom and darkness. The air moaned on every side as if it was calling out "death, death.". . . Within the ship nothing was heard but sighs, groans, moans, and prayers to God for mercy. . . . Between the decks it seemed as if the evil spirits were busy, so great was the noise made by the things which were floating about and bumping against each other, and crashing from side to side, so that those who went down below fancied that they beheld a likeness of the last Judgement.[19]

The physical activity of working the pumps or bailing with buckets perhaps gave the passengers and crew a sense of control over their own destiny, or maybe it just distracted them from their fears for a little while. The common and present danger of sinking was a powerful eliminator of social distinctions.

> Our governor . . . had caused the whole company, about 140, besides women, to be equally divided into three parts and, opening the ship in three places . . . appointed each man where to attend; and thereunto every man came duly upon his watch, took the bucket or pump for one hour, and rested another. Then men

> might be seen to labor, I may well say, for life; and the better sort, even our governor and admiral themselves, not refusing their turn and to spell each the other, to give example to other. The common sort . . . kept their eye waking and their thoughts and hands working with tired bodies and wasted spirits three days and four nights, destitute of outward comfort and desperate of any deliverance, testifying how mutually willing they were yet by labor to keep each other from drowning, albeit each one drowned whilst he labored.[20]

As the danger increased, measures were taken to improve the stability of the ship. The masts could be cut down if the ship had gone on her beam ends or guns jettisoned, but most commonly, the cargo was thrown overboard. Under these circumstances, the socioeconomic hierarchy was still respected: the first things to be cast out were the belongings of the poor, the soldiers, and the common seamen. Next came the richer cargoes of the merchants, officers, and nobles. If all else failed, Crown and Church property were the last to be discarded.[21]

When the people saw that they could no longer hope to save the ship or their lives, some would frantically try to save their souls. "Those on board . . . stormed around the priests. . . . In their confused animal madness all of them wanted to confess at once and began to speak their sins so loudly that they were heard by all. . . . As the desperation grew, one crazed individual, seeing that he could not soon receive confession, began to shout his sins above the tumult of everyone, sins that were so enormous a priest put his hand over the man's mouth."[22]

Others desperately tried to find a place in one of the few ship's lifeboats. Still others resigned themselves to their fate. "The people . . . now seeing their efforts useless, many of them burst into tears and wept like children. . . . Some appeared perfectly resigned, went to their hammocks and desired their messmates to lash them in . . . but the most predominant idea was that of putting on their best and cleanest clothes."[23]

Just as there are many people, so are there many reactions. One person's disaster can be another's good fortune.

> In the midst of despair and confusion, a slave was so delighted to see himself separated from his master—who was one of the first to get into the lifeboat—that he made no attempt to save himself. He ate sweets from the neglected barrels of provisions, swam gaily

around the broken sides of the *Santiago* and sang out that he was a free man. For a few hours the slave, of his own free will, was able to trade his chances of survival for the sudden privileges of a freedom created by disaster.[24]

The pumps, supplemented in severe cases with bailing buckets, were the last defense, hope, and salvation of the lives on board. If one were on a sinking ship and had the choice of having working sails and rudder but no pumps or of having no sails or rudder but working pumps, which would one choose? The pumps were the most important pieces of equipment on a ship.

2

The Construction of Wooden Tubes

Wooden pumps for ships differed little from those made for land use: the major distinction lay in the shape of the exterior and the lower end. The procedures for fashioning the tubes were precisely the same. Elm was the most popular wood used because of its durability in a wet environment.[1] Other woods with similar properties, such as larch, beech, and alder, were also used.[2] The pump tube from the American privateer *Defence*, sunk near Castine, Maine, in 1779, is believed to be of pine.[3]

The first task in making a wooden pump tube was to find a straight tree trunk free of knots and side branches. Such a log could be difficult to locate because there was always the possibility that some branches had been trimmed from the tree in the past and that their knots had been overgrown with new bark.[4] For this reason, trimming the outside of the log to its final shape was sometimes done first.

The log was set up on a platform or supports of some kind and fastened thereto (Fig. 1). It could be secured by rope or chain, but a common method was to use "dogs." Dogs were pieces of iron ca. 30.5–45.7 cm (12–18 in) long with the ends bent at right angles and the tips sharpened, much like large staples. One tip was driven into the log and the other into the support to bind firmly the one to the other.[5] Accounts kept during construction of the frigate *Essex*, built 1798–99, show that copper dogs were specially made for the boring of the ship's pump tubes.[6]

The pump makers marked with string longitudinal guidelines down the length of the log at the top and along the sides; they then

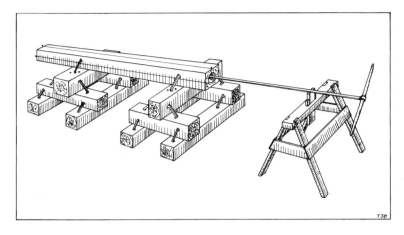

Fig. 1. Boring a wooden tube. The log is placed upon a platform and fastened to it with dogs. The auger shaft is supported by the auger stool.

located the centers of each end using plumb bobs and squares. With the auger bit at the center of one end of the log, the auger shaft was aligned with the strings at the top and sides to ensure that the hole would be in the center of the tube.[7]

In order to achieve the desired bore size, a set of from three to six or more "shell" augers (bits) of graduated diameters was used to cut successively larger holes. The bit that cut the pilot hole through the center of the log was known as a quill bit,[8] nose auger, pod auger, or split-nose auger.[9] It was long with straight, parallel sides; the cutting edge was at one extremity and perpendicular to the shaft (Fig. 2). Since shell augers did not have pointed tips, a small hole or depression had to be made with a chisel or gouge so that the first bit could bite into the wood. Screw-pointed augers were not used because they tend to follow the grain of the wood. If a screw auger hit an undetected knot or shift in the grain, it would veer from its proper course and exit somewhere along the side of the log.[10]

After the pilot hole had been drilled completely through, the diameter of the bore was increased with successively larger bits. These were tapered shell augers, each of which had a sharpened cutting edge along one side as well as at the end. This type of tool is more efficient than a shell bit with a gougelike end, especially in the larger bit sizes.[11] A strap of iron or a piece of wood was often bound to the larger bits, opposite the cutting edge, through two holes provided for that purpose (Fig. 2d). In this manner the bore diameter could be made larger than that bit but still smaller than the next one in the graduated series.[12]

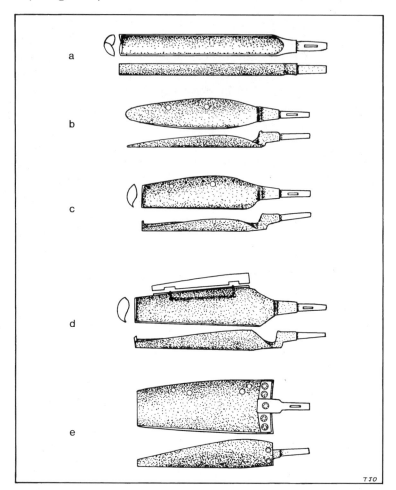

Fig. 2. Shell augers and bits: a, *quill auger;* b, *spoon auger;* c, *third bit;* d, *fourth bit with iron strap attached;* e, *sixth bit. (After O'Sullivan 1969: 104)*

The auger shaft was ca. 3.66–4.57 m (12–15 ft) long, with a wooden handle ca. 91.4 cm (3 ft) or more in length at one end. At the other end was a square socket into which the bits were fastened with a key that passed through the shaft and bit. An auger stool, essentially a stable platform with an adjustable rest, was used to support the auger shaft. It was positioned and the shaft adjusted so that the latter lined up with the longitudinal guide strings on the log (Fig. 1). Two men, facing each other with the handle between them, twisted the auger in quarter turns while a third man, with his back to the log, pulled the bit against the log.[13]

Because the bit quickly filled with shavings, it was often turned in reverse, withdrawn, and cleared. A hooked bar might also be kept on hand for this purpose. When the handle drew near the stool, the next bit was fitted to the shaft and so on until the hole reached the required diameter. Then the first bit was refitted, the stool was moved forward, the alignment of the shaft was checked, and boring began once more.[14] An auger shaft in the Mercer Museum in Doylestown, Pennsylvania, displays markings in feet along the shaft that enabled workers to determine at a glance what progress they had made.

Tubes over ca. 4.57 m (15 ft) in length were bored from both ends. It was not necessary for the holes to meet perfectly, and the diameter of the hole could differ from the top to the bottom of the tube as well. Such imperfections were actually desirable. The variation in bore diameter served to compensate for the volume taken up by the pump rod.[15] The offset or the different diameters that resulted from boring from both ends allowed the lower valve of the common pump to seat conveniently and safely, without danger of falling down the tube.

Although most tubes were made by hand, a few machines were designed to do the work. Leonardo da Vinci drew a pipe-boring machine in 1500,[16] and Belidor illustrated one powered by a water wheel.[17] A mill was set up to bore pine tubes in England in the 1770s; another, operating in London, produced tubes of elm.[18]

Pumps used on land were often made of two and sometimes three short tube sections instead of one long one. A ship's pump was usually a single long tube, although two sections were occasionally used in the early years of the sixteenth century. Two segments of the ship's single pump tube have been recovered from the wreck of the *San Juan*, a Basque whaling ship that sank in Red Bay, Labrador, in 1565. They do not now fit together because the wood has eroded, nor have growth-ring analyses of the two segments proved conclusive as to how many sections the tube comprised.[19] From the wreck of the *Machault*, a French frigate that sank in 1760 during a battle on the Restigouche River in Canada, the tubes of three different pumps were recovered intact. The overall length of one was 7.6 m (ca. 25 ft). On most shipwrecks, all that remains of the pumps are the lower ends of the tubes.

Ships generally did not carry the specialized tools required for the fabrication of tubes, although a spare tube may have been carried for emergencies.[20] A method of making a pump tube without the proper bits was described in 1687 by William Dampier:

> And our Pumps being faulty, and not serviceable, they did cut a
> Tree to make a Pump. They first squared it, then sawed it in the

middle, and then hollowed each side exactly. The two hollow sides were made big enough to contain a Pump-box in the midst of them both, when they were joined together: and it required their utmost skill to close them exactly to the making a tight Cylinder for the Pump-box; being unaccustomed to such Work. We learnt this way of pump making from the Spaniards; who make their Pumps that they use in their Ships in the South-Seas after this manner.[21]

It is evident that this method, born of necessity, produced serviceable pump tubes even when proper tools were not available. Two examples of such tubes were found on a Spanish wreck near Port Royal, Honduras, in the early 1970s. Although the wreck was tentatively dated to 1515, based on olive jars, the ship also carried "straight-sided rum bottles" that are probably of eighteenth-century date.[22]

A tube constructed from planks was another alternative to a bored tree trunk. The planks were shaped so their edges fit snugly, the seams were caulked, and the entire assembly was secured at intervals with straps. Not only the tube but also the upper and lower valve boxes were square. Such tubes were used to some extent by the U.S. Navy as late as 1820.[23]

William Hutchinson invented a type of diaphragm pump that included square tubes built of planks.[24] He referred to this type of tube as though it were nothing new. It was certainly simple enough to have been in both naval and merchant service for a long time, but the necessity of caulking the four seams probably kept it from being used on a permanent basis because of the additional maintenance required. A pump made of elm staves firmly bound together with ash hoops and lined with cowhides was devised by James Brindley (1757–1812).[25] Similarly, such a tube probably did not become popular because of its complicated construction.

The bottoms of tubes used in land wells were treated differently from those found on ships. For application on land, the lower extremity of the bore was plugged, and small holes were drilled into the sides a short distance from the bottom. In this way, sediment and other material at the bottom of the well were not sucked up the tube.[26] This was a necessary precaution when drinking water was involved, but for ships' pumps the bore was left open so that the water level in the hold could be reduced as much as possible.

For a ship's pump, the heel of the tube would be fashioned to fit securely in the bottom of the hull. The specific way in which this was done varied with the different types but was also dependent on the whims of the carpenters who installed it. Therefore, the basic manner of seating each kind of pump tube is addressed in its appropriate chapter.

The methods and tools used in boring wooden tubes for pumps changed little over the centuries. Tubes for ships' pumps were built in the same way as those for land pumps. The tools conformed to the same basic pattern (auger shaft, bit set, stool, platform, etc.) and were used in the same way. Alternatives to a bored log existed in tubes made of planks and in hollowed halves of a log, but these were not as durable as the former.

3

The Burr Pump

The burr pump is an extremely simple machine and may have been in use well before the Renaissance. In addition to the tube, its components included the foot (lower) valve and the spear (pump rod) with the burr (upper) valve attached (Fig. 3). The foot valve also acted as the base of the pump in some cases. In this capacity, it consisted of a short piece of wood fashioned into a tube with a valve claque at its top.[1] The upper valve was a thick piece of wood, often in the shape of a truncated cone, called the burr. A wooden pole of sufficient length was attached to the cone's base, and a piece of leather was fastened around the cone. The leather extended the wooden cone upward beyond its base so that the base diameter of the leather cone equaled the tube's bore diameter.[2]

When the pump was primed with water and the rod or pole was moved up and down in the tube, the leather cone closed on the downstroke, opened on the upstroke (much like an umbrella), and lifted the water above itself. At the same time, water was drawn up through the foot valve. To reinforce the leather cone, strips of leather were sewn to its base and fastened to the spear.[3]

The simplest way to impart an up-and-down motion to the spear was to grasp it by hand and move it up and down, perhaps by means of a handle in the form of a short crosspiece fastened to the top of the spear. Alternatively, a standard pump brake, which was a lever bar mounted on a fulcrum within a yoke, could be set on or near the pump head.[4] Manwayring and Boteler, however, both describe a different system used on ships in the early seventeenth century. Two

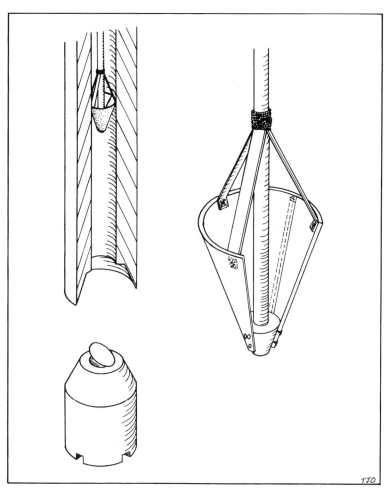

Fig. 3. The parts of a burr pump. Left: *tube with burr valve in place and foot valve below.* Right: *detail of burr valve.*

men stood facing each other over the pump head and pushed the spear down; six to ten men then hauled it up with a rope seized to the middle of the spear.[5] Portions of the spear, tube, burr valve, and foot valve were recovered from the wreck of the Basque whaling ship *San Juan* (1565). The foot valve, essentially a stepped cylinder that also served as the pump's base, still exhibits a valve claque made of six disks of leather fastened together (Fig. 4). Two of the leather disks had broad tails that were nailed to the valve in a recess cut for that purpose. An eroded foot valve was found in place on the Studland Bay wreck, a Spanish vessel of the late fifteenth or early sixteenth centuries.[6] The tube fit over the top portion of the valve and rested on the step. The

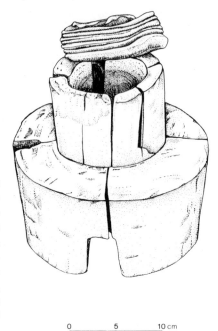

Fig. 4. The foot valve and leather valve claque from the Basque whaler San Juan *(1565). (Drawing by S. Laurie-Bourque, Canadian Parks Service)*

San Juan's burr valve differed from the standard design in that instead of a leather cone attached to the spear, it consisted of twenty-one leather disks of two sizes affixed to the end of the spear with an iron pin.[7] Smaller underlying disks supported the center of the stack of larger ones (Fig. 5).

Although no other evidence of burr pumps has been discovered on sixteenth-century wreck sites, there are similarities among the pump seatings found on the wreck of the *San Juan* and several others (Fig. 6). The seating for the *San Juan*'s pump was created by cutting a semicircular hole into the after port side of the main mast step, which was an enlarged portion of the keelson. A floor timber beneath the keelson was partly cut away for the pump as well.[8] A similar seating was found on a wreck dated to the sixteenth century discovered near Rye, Sussex (England).[9] A semicircular hole is let into the side of the expanded portion of the keelson ca. 28 cm (11 in) aft of the mast step. The Cattewater wreck in Plymouth, England, also dating to the sixteenth century, exhibits a tapering elliptical hole cut into the center of the mast step.[10] The wreck of Iberian origin at Highborn Cay in the Bahamas displays two semicircular holes of some 27 cm (ca. 10.5 in) in diameter in the port side of the keelson.[11]

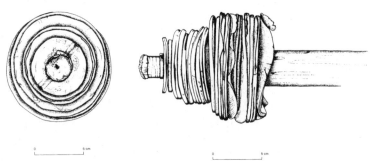

Fig. 5. The pump spear and leather valve from the San Juan. *(Drawing by S. Laurie-Bourque, Canadian Parks Service)*

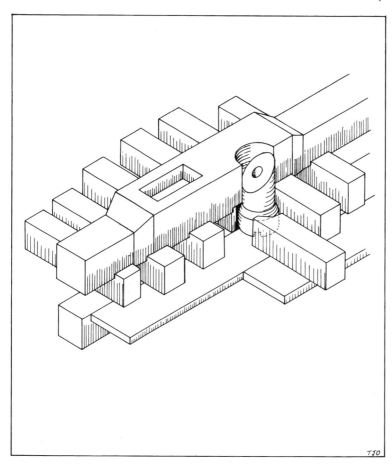

Fig. 6. A typical sixteenth-century pump seating. Portions of the mast step and a frame have been cut away to receive the foot valve.

Whether there had been two pumps or a repositioning of a single pump has not been determined. The Emanuel Point wreck (1559), a vessel of the Tristan de Luna expedition to Pensacola, Florida, had a sump on each side of the mast step, again with a semicircular cut carved into the keelson.[12] The Western Ledge Reef wreck (Spanish, sixteenth century) from Bermuda exhibited a variation in this pattern: there was a flat indentation in the port side of the keelson instead of a semicircular cut.[13] Although there is a similarity between the mast step construction and pump seatings of these ships, we cannot associate a burr pump (as opposed to a common pump) with such a seating because it is the upper valve of the pump that determines which of the two it is.

Peter Waddell notes that the seating hole on the *San Juan* was cut in a crude manner, which suggests to him that the placement of

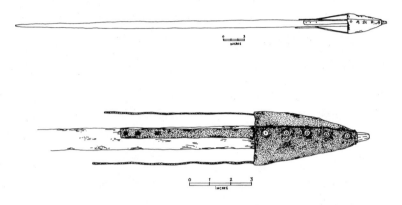

Fig. 7. The burr valve and spear from the mid-nineteenth-century wreck of a canal sloop found near Isle La Motte, Lake Champlain. (Drawing by Kevin Crisman)

Fig. 8. Burr valve and spear from the wreck of the canal sloop near Isle La Motte, Lake Champlain. (Photo by Kevin Crisman)

the pump was an afterthought. He also points out the contradiction of enlarging the keelson to provide a stronger mast step and then cutting into it to create the pump seating. This was done so that the combined weight of the tube and the water column was borne partly by the keel and thus not by the planking alone.[14]

When and where the burr pump was first used is not certain. It is cruder in design than the common (or suction) pump, which suggests it was a predecessor of the latter. Georg Agricola's *De Re Metallica*[15] is the earliest work I have found that illustrates a burr pump, which was said to be inferior to the common pump.

By the beginning of the seventeenth century, the burr pump had fallen out of general use. Marine dictionaries of the period mention it and state that it was no longer in service on English ships but that it could be found on Dutch and Flemish vessels. The latter had broad, flat bottoms, and their pumps were placed at the ships' sides in order to expel the water that collected at the turn of the bilge when the ship heeled.[16] Two burr pumps were reported on the Lelystad Beurtschip, a flat-bottomed Dutch vessel built before 1600 that sank in the 1620s.[17] The burr pump did occasionally appear in later times, however, because it was a simple machine and was easy to construct.[18] A burr valve and spear were recovered from the wreck of a canal sloop found near Isle La Motte in Lake Champlain and dated to the 1840s or 1850s (Figs. 7, 8). The leather cone of a burr valve was also discovered on the wreck of the *General Butler* (built 1860, lost 1876), a canal schooner that sank in the lake as well.[19] These later burr pumps were modified in that the large foot valve was replaced by the lower valve arrangement of the common pump.

Both Boteler and Manwayring state that the burr pump drew much more water and was less labor-intensive than the suction pump of the time.[20] If this was true, one wonders why the former went out of service. Perhaps it was after these two works were published that the common pump gained prominence. But the burr pump's separate foot valve surely disappeared because of the difficulty of servicing it. In order to replace, repair, or clear the leather flapper valve, the entire tube had to be lifted off the base. This would have been difficult at sea within the close confines of the well, especially during rough weather or after a few feet of water had collected in the hold. The common pump had definite advantages in this regard; in general, advances in technology permitted improvement of this pump so that it far surpassed the burr type.

4

Common or "Suction" Pumps

> *The common pump is so generally understood, that it hardly requires any description. It is a long wooden tube whose lower end rests upon the ship's bottom, between the timbers, in an apartment called the* well, inclosed *[sic] for this purpose near the middle of the ship's length.*
>
> *This pump is managed by means of the brake, and the two boxes, or pistons. Near the middle of the tube, in the chamber of the pump is fixed the lower-box, which is furnished with a staple, by which it may at any time be hooked and drawn up, in order to examine it. To the upper-box is fixed a long bar of iron, called the spear, whose upper-end is fastened to the end of the brake, by means of an iron bolt passing through both. At a small distance from this bolt the brake is confined by another bolt between two cheeks or ears, fixed perpendicularly on top of the pump. Thus the brake acts upon the spear as a lever, whose fulcrum is the bolt between the two cheeks, and discharges the water by means of valves, or clappers fixed on the upper and lower boxes.*
>
> *William Falconer,*
> An Universal Dictionary of the Marine

The earliest representation of a common, or suction, pump is a drawing found in the 1431 notebook of Mariano Jacopo Taccola, an Italian engineer. The drawing shows a tube only a few feet in length, which has been interpreted as proof that the capabilities and limitations of the pump had not yet been discovered.[1] Drawings of the pump appear in later engineering texts, and by the second quarter of the sixteenth century, the suction pump had found practical application in the mining industry in Germany. The use of suction pumps on ships, therefore, probably began sometime in the late fifteenth or early sixteenth century.

A predecessor of the common pump is the force pump known from Greek and Roman times.[2] In a force pump, the piston is solid and is situated between two valves that permit the water to flow in only one direction. In the harbor of Kalmar, Sweden, part of a pump tube was found on Find V, a sixteen-meter-long boat of ca. 1500. It is reconstructed as part of a force pump. The tube rested on a base that housed the valves, and a second tube conducted the effluent to a trough on deck.[3] The Ringaren wreck (ca. 1542) from southern Sweden has a pump that is reconstructed as a force pump.[4] The implication is that the ancient force pump was still in use around 1500 in Scandinavia and that the common pump may not yet have reached this region.

The height to which a common pump can raise water by suction is governed by the physical laws of barometric pressure. A rule of thumb is that suction can lift a column of water to a height in feet equal to the barometric pressure as expressed in inches of mercury. That is, at a barometric pressure of 30 inches (ca. 76.2 cm) of mercury, it is theoretically possible to raise a column of water to a height of 30 ft (ca. 9.15 m). Because of friction and the inefficiency of the pumps, however, the height was usually closer to 28 ft (ca. 8.54 m). This "critical distance" is measured from the surface of the water to the claque of the upper valve at the top of its stroke (Fig. 9). The length of the pump tube was determined by the depth of the ship's hull. Tubes longer than 28 ft were entirely possible, because the upper valves could be placed some distance down the tube, usually near the midpoint, thereby bringing the critical distance to within operating limits.

A pump had to be primed when there was no column of water in the tube. Water was poured down the tube to cover the upper valve, thus sealing off the lower part of the tube from air. As the piston worked, the atmospheric pressure decreased within the tube. The water then rose through the lower valve because of the greater at-

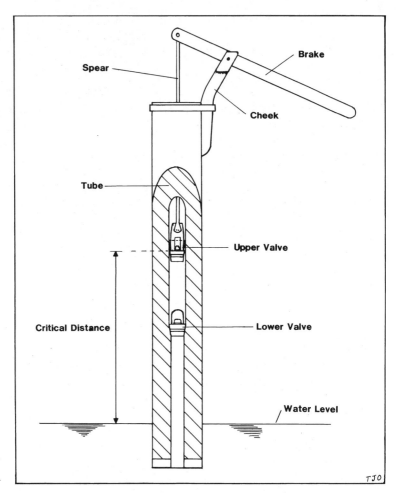

Fig. 9. Schematic representation of a common pump.

mospheric pressure forcing the water down outside the tube. This was repeated until all the air was removed from the tube and water rose through the upper-valve claque.

The shapes and proportions of the different forms of valves were well known to pump makers; as long as the valves were made of wood, they retained quite unvarying and recognizable shapes. In fact, a comparison of eighteenth-century pump valves from the *San José* (1733); the *Machault* (1760); the *Defence* (1779); the Pump Museum in Saco, Maine; and the Mercer Museum in Doylestown, Pennsylvania, reveals them to be identical in overall shape and design to wooden valves for land pumps produced by Irish pump makers in the first part of the twentieth century.[5]

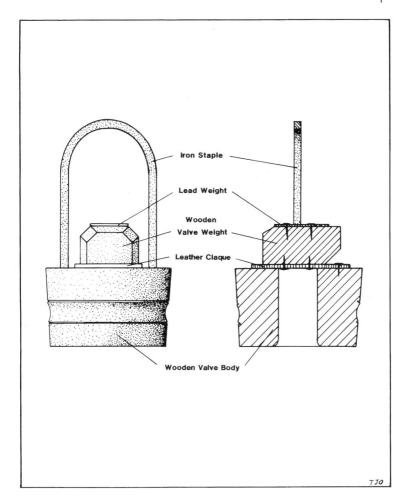

Fig. 10. Schematic representation of a common pump's lower-valve body. The vertical hole was sometimes round, sometimes oval.

Generally, valves were turned on a lathe and made from elm (Fig. 10). Lines were scribed around the body where significant features began or ended. The lower valve's thickness was equal to or a little less than its diameter. The large hole in the center, through which the water passed, was usually oval but sometimes round. Holes on each side of the central hole received the ends of a staple, and these ends were secured by riveting them over roves at the bottom of the box. A staple enabled the lower box to be retrieved for inspection and repair. For this purpose, a long pole with a hook on the end was always kept on board. This feature gave the common pump a decided advantage over the burr pump because only the lower valve, not the entire tube, had to be removed in order to replace the valve leather.

Fig. 11. *The body of a lower piston valve from the* Machault *(1760) showing the grooves for a basal-ring type stirrup. (Photo by author, courtesy Canadian Parks Service)*

Another type of staple in a French illustration was connected to a basal ring, both of which fit into grooves on the exterior of the wooden valve.[6] An example of this kind of valve, from the *Machault* (1760), is in the Canadian Parks Service collection (Fig. 11). Its valve claque (a piece of leather large enough to cover the central hole, with a small block of wood and/or a piece of lead attached to its upper surface to act as a weight) was fastened along a top edge of the valve body with tacks. A groove encircled the outside of the valve body, and here a wrapping of flax, yarn, wool, or the like ensured a watertight seal around the box.[7]

An upper valve (Fig. 12) consisted of the body and the valve claque. A circular hole cut through the center from the bottom intersected a large rectangular hole sliced through the diameter. The valve claque was placed atop the circular center hole. Just below the level of the claque an indentation was cut in the body, where a piece of leather was wrapped around it. The lower edge of the leather rested on the ledge created by the indentation, and the butted ends were tacked to the side of the body. The leather was intended to prevent water from leaking past the valve. Farther down on the body, a short distance from the base, a groove was cut around the circumference. Although similar to the groove on the lower box, no source mentions a wrapping around the body at this point.

Common or "Suction" Pumps 27

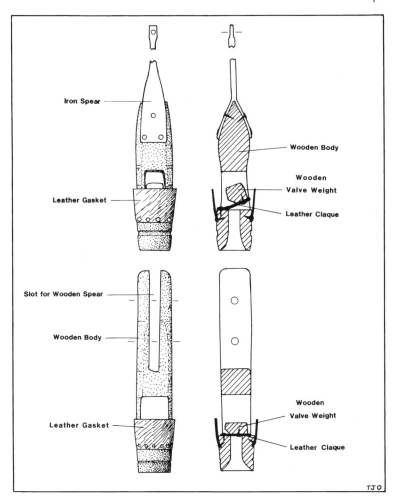

Fig. 12. Renderings of two types of upper valves. Top: *shaped to fit a forked iron spear.* Bottom: *slotted for attachment to a wooden or iron spear. (Both after O'Sullivan 1969: 112)*

The upper portion of an upper-valve body was shaped according to the type of spear used. The top of the body took the form of a wedge if a metal spear with a forked end was to be affixed. The flattened tongs of the fork fit over the wedge and were nailed in place (Fig. 12, *top*). For a wooden or metal spear whose extremity was flattened or eye-shaped, a slot was cut from the top of the valve body to a short distance above the horizontal, rectangular hole. Holes were then drilled across the slot, through which the spear was bolted or pegged (Fig. 12, *bottom*). When the spears were made of wood, fir or pine was preferred.[8]

Complete examples of all the above-mentioned types of upper

Fig. 13. Two upper-valve bodies, a leather gasket, and two lower-valve bodies, all from the Machault *(1760). (Photo by author, courtesy Canadian Parks Service)*

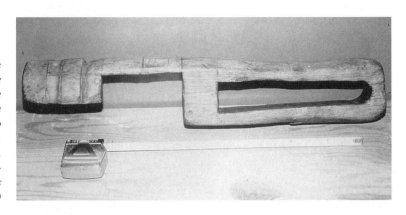

Fig. 14. An upper-valve body from the Machault with a slot in the top to accept the spear. (Photo by author, courtesy Canadian Parks Service)

and lower valves, made of ash and elm, were recovered from the *Machault* (1760) (Figs. 13, 14). They resemble very closely the examples given by John O'Sullivan, who recorded the process of pump making as practiced in the first part of the twentieth century.[9] A few were probably spare valves because there are no nail holes where the valve claques would have been tacked down. A fragment of a lower valve was discovered on the wreck of the *Defence* (1779), and several lower valves were recovered from the *San José y las Ánimas*, a ship of the Spanish Plate Fleet that was lost off the Florida coast in 1733.[10] Piston valves from the Swedish shipwrecks at Jutholmen (seventeenth century) and Älvsnabben (ca. 1730), although slightly different in appearance, display all of the major features described.[11] Other valves from land pumps of the late nineteenth century are in the Mercer Museum in Doylestown, Pennsylvania, and in the Pump Museum in Saco, Maine. The similarities shared by all these valves, coming from such different locales and periods, are striking.

The "brake" or handle of a common pump was usually a wooden bar bolted perpendicularly through supports or "cheeks" fashioned out of the tube itself or attached as a separate unit. Examples of the latter include those on the *Vasa*'s (1628) pump, a decoratively carved cheek with a tenon from the Jutholmen wreck, and a cheek piece of iron on the *Victory*.[12] In photographs of the *Scourge* of 1813,[13] the cheeks are quite evident and are part of the tube; the top of the spear is visible next to the cutlass in the port pump. The brakes are not mounted, but the square timber protruding from the bore of the starboard tube may be one of them. On the wreck of the *Machault*, one brake was found in the bore of a pump, next to the spear. This is a logical and handy place to stow the brake. The spear could not have fallen very far down the tube because of the lower valve, and the brake could have been wedged between the spear and the tube wall.

Another form of brake was a long lever arm temporarily lashed near its midpoint to the mast, above the pumps. A rope tied to one end of the arm led to the spear, and a number of lines were attached to the opposite end so that as many men as could grasp the lines could work the pump. The spear was not forced back down by hand, since weights were attached to its top to do the job. Jean Boudriot shows cannonballs (probably imperfect and therefore useless as shot) bound in iron straps with an eye for attachment to the spear.[14] An object that could be one of these overhead lever arms was found on the seventeenth-century Jutholmen wreck.[15]

A more advanced brake mechanism is the bronze quadrant of the Taylor's pump from HMS *Pandora* (1791). A section of the outside edge of the quadrant was toothed, which engaged a rack on the end of the pump spear. This allowed a more vertical movement of the spear and reduced wear on the leather gaskets.[16]

Water usually left the pump tube through a hole in its side near the top. Because the water in the bilge had its own distinctive odor, it was common to attach a conduit to the opening that guided the bilge water to a scupper or directly over the side instead of allowing it to spill out onto the decks. This conduit, called a dale, was made of canvas or planks and could be removed when not in use. Lead pipes were sometimes employed as permanent dales.

As mentioned above, the lower extremities of ships' pumps were fashioned in a manner different from those used on land. The bore was left open at the bottom of the tube, and, to allow water to enter the bore at the base, four channels were carved along radii to the center. The heights and widths of these cuts varied from pump to pump.

In order to anchor the heel of the pump firmly in the bottom of the ship, one or more facets were created at the bottom end to permit the tube to rest between the floor timbers or against the keelson (Fig. 15). Examples of this technique have been found on the wrecks of the *Defence* (1779), the *Machault* (1760), and at Trinity Cove (mid-eighteenth century).[17] In similar fashion, a circular hole a few inches deep could be cut into the top of a floor timber to receive the heel of the tube. The two suction pumps of HMS *Charon* (1781), a fifth-rate, forty-four-gun ship, were seated this way (see Fig. 45). In contrast, the common pumps of the *Santo António de Tanna*, a Portuguese frigate that was sunk in the harbor of Mombasa, Kenya, in 1697, were set between frames through holes in the ceiling planking.[18]

One of the most frequent difficulties with common pumps was that debris from the bilge could be drawn up into the tube and foul one or both valves. To guard against this occurrence, a perforated sheet of lead or copper, called a sieve, was sometimes fitted over the lower extremity of the pump. The sides of the sieve were bent up around the outside of the tube and nailed with broad-headed tacks. Lines delineating the channel cuts, scribed before nailing, are sometimes found on the interior surfaces of sieves. Within these lines a number of holes pierce the lead or copper; tools used to pierce the metal varied from hot pokers to gouges.

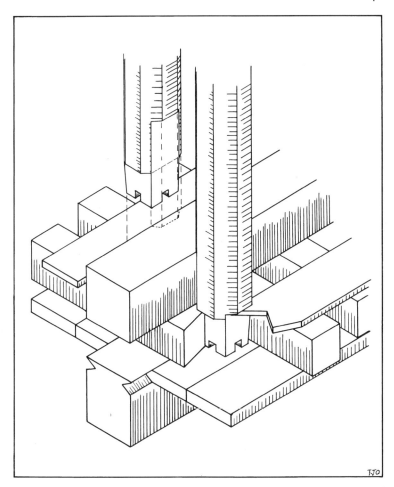

Fig. 15. A typical common pump seated between the frames. The lower ends of the tubes have been shaped to fit in their assigned positions.

The various sieves recovered from wreck sites to date are all made of lead. The earliest is from the *Sea Venture* (1609), which wrecked in Bermuda while carrying English settlers to Jamestown.[19] Of the two recovered from the *Machault*, one is a thick-sided box (three sides and a bottom), the other a thinner sheet fastened by means of small tabs that overlap the exterior of the tube. The bottom ends of two tubes from *El Nuevo Constante*, a Spanish ship that wrecked on the Louisiana coast in 1766, were found with the sieves still attached (Fig. 16).[20] A third pump, possibly a spare, is suggested by a third and much larger sieve from the site.

It was not always refuse that clogged a pump. Certain items of cargo or ship's stores, when they became wet, could also wreak havoc with the pumps. The Portuguese, on their return voyages from India, had continual problems with pepper choking the pumps. When the peppercorns became saturated, they swelled, burst the sacks in which they were stored, and migrated to all parts of the hold, including the pumps and the well. When this happened during the 1589 voyage of the *Saõ Thomé*, sheet tin was used to prevent a recurrence. Although it is not known how the tin was used, it most probably served as a sieve at the bottom end of the tube.[21] Coriander, an herb whose aromatic seeds are about the size of peppercorns, was found between the valves in one of the preserved tubes from the *Machault*.[22] During a voyage of the *Sea Venture*, the ship faced a desperate situation in a storm when the pumps became congested, "bringing up whole and continual biscuit."[23] The loss of the English third-rate *Centaur* (1782) was in part caused by its load of coal. Rising water allowed the coal

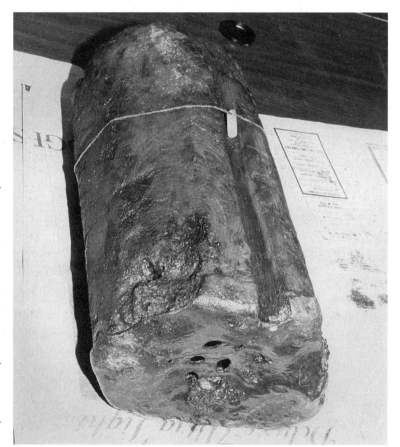

Fig. 16. The lower end of one of the asymmetrical-shaped pump tubes from El Nuevo Constante *(1766). The lead sieve is still in place, and a groove of unknown purpose runs down the outside of the tube. (Photo by author, courtesy Louisiana Division of Archaeology)*

Common or "Suction" Pumps   33

to shift within the hold. Once the lumps reached the pump well, they repeatedly clogged the common pumps and damaged the chain and leathers of the chain pumps.[24]

An unusual feature of tubes found on some shipwrecks is a semicircular groove extending down their exterior, the function of which has not been explained. Grooves have been found on the one tube from the Trinity Cove wreck (Newfoundland)[25] and on two tubes and an extra lead sieve (as mentioned above) from *El Nuevo Constante* (1766) (Figs. 16, 17).[26] From the latter ship, a lead sieve with a small semicircular cut in one corner suggests the existence of a third tube with an exterior groove. A complete pump tube with the groove running the entire length and passing under several iron straps was found on the wreck of the *Sydney Cove* (1797).[27] Each of two wooden common pump tubes from the *De Braak* (1798) had a U-shaped metal channel attached to its outer surface.[28] In the absence of literary evidence, we can only speculate about the purpose of these grooves. The most logical idea is that they provided a space in which a weight or rod attached to a string could be lowered to measure how much water was in the hold.

Fig. 17. Common-pump sieve from El Nuevo Constante (1766). Note the large holes, nail holes, scribe lines, and semicircular cut in the top right corner. (Photo by author, courtesy Louisiana Division of Archaeology)

As technology advanced, different metals became available to improve the durability and efficiency of the pumps. The first extant record of a metal pump (that is, a pump whose valve bodies, tube, or some portion of the tube is made of metal) was made by Diego Rivero, cosmographer and maker of instruments and maps for the Spanish Crown. He invented one in 1526 for use on ships bound for the Indies, claiming his pump was lighter than those of wood and that it could move many times the amount of water a wooden pump could.[29] There is no description of the mechanism or of the metal used to make this pump, so a positive identification of the pump type is impossible. It may, however, be what Cesáreo Fernández Duro referred to when he stated that the Spanish invented copper pumps in the first half of the sixteenth century.[30] Because of the date and the use of metal, Rivero's pump was most likely a suction pump. By 1556, Georg Agricola was recommending that pump valves be made of iron, copper, or brass.[31]

Employment of metals in pumps, especially iron and copper, began in the sixteenth century, increasing gradually through the nineteenth century. During the sixteenth and seventeenth centuries, however, the manufacture of metal machinery was less cost-effective than was the production of metal ordnance and luxury goods. Therefore, wood remained the basic construction material; metals were used only where strength and durability were essential,[32] and their application grew rather slowly. Yet lead was easily obtained and worked. Lead tubes or conduits were used in water-supply systems in the sixteenth century,[33] and it appears that the material was being utilized in ships' pumps early in the sixteenth century, before iron and copper.

Two identical objects recovered from the Molasses Reef wreck, an early sixteenth-century Spanish shipwreck discovered in the Turks and Caicos Islands (British Virgin Islands), are elements of simple piston valves (Fig. 18).[34] They look very much like wheel hubs, which they were first suggested to be by the

Fig. 18. One of the two lead disks recovered from the Molasses Reef wreck. (Photo by author. Source: Turks & Caicos National Museum)

treasure hunters who retained one of them. The objects are made of lead, however, and would not have withstood such stresses. Each is a disk about 1 cm thick with a raised reinforcing collar around a central hole that is surrounded by seven smaller holes. A horizontal hole through the collar allowed it to be secured to the rod, axle, spear, or whatever passed through it. Such disks would have been quite satisfactory piston valve components and could represent an early form of upper valve.[35] The valve claque would have been a circular piece of leather with a central opening for the spear; tabs left when the opening was cut could have provided a means of lashing the leather in place (Fig. 19). The small holes in the disk would have allowed water to pass up through the valve on the downstroke, and the leather would have closed them on the spear's upstroke. The disk is similar to the valve described by Agricola (Figs. 20, 21):

Fig. 19. Common pump upper valve from the Molasses Reef wreck, as reconstructed by the author. (Photo by author. Source: Turks & Caicos National Museum)

> Or else an iron disc one digit thick is used, or one of wood six digits thick, each of which is far superior to the shoe [the burr valve leather]. The disc is fixed by an iron key which penetrates through the bottom of the piston-rod, or it is screwed on to the rod; it is round, with its upper part protected by a cover, and has five or six openings, either round or oval, which taken together present a star-like appearance; the disc has the same diameter as the inside of the pipe, so that it can be just drawn up and down in it.[36]

Tubes of lead, the first metal used for that part of a ship's pump, were made to any diameter by cutting the width of a lead sheet to the desired tube circumference, rolling the sheet so the sides met, and soldering the seam.[37] Such tubes were durable, did not rust, and were not prone to splitting as wooden ones sometimes were.

An early example of a lead common pump comes from the wreck of the *San José* (1733), possibly an English-built ship purchased by the Spanish. It consists of a lead reservoir box with no top, a dale,

Fig. 20. A woodcut from Agricola's De Re Metallica *(1556). Various pump parts, including what are probably metal disks similar to the two recovered from the Molasses Reef wreck, are clearly evident in the left foreground.*

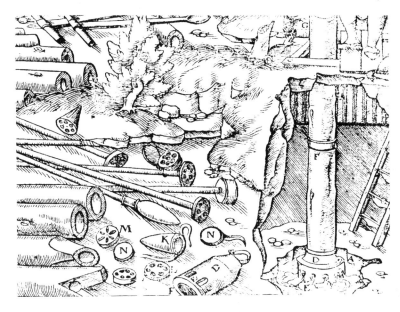

Fig. 21. Detail of fig. 20.

and a piston tube (Figs. 22–24). The bottom of the reservoir box was soldered to a single sheet of lead wrapped around it that formed the sides, and a lead pipe attached near the bottom of the box's back served as a dale. The top of the piston tube was soldered to the underside of the box's bottom. A bronze gravity valve (i.e., a valve that seats itself by gravity and is not attached to the sides of the tube or valve seating in any way) was set permanently into the tapered lower end of the piston tube. This tapering extremity probably fit into a wooden tube.

Two lead pump tubes in the Mariners' Museum, RS 2 and RS 3 (Fig. 25), were recovered from the York River at Yorktown, Virginia. This is the site of General Washington's victory over Lord Cornwallis in 1781 and the final resting place of a number of ships of the British supply fleet. The tubes were salvaged in the 1930s during a joint project conducted by the National Parks Service and the Mariners' Museum[38] and bear some resemblance to the one from the *San José*. They are each attached to an open reservoir box, with a drainage port at one side, mounted on top of a piston tube. A photograph of these tubes immediately after their recovery shows two seating flanges and one reservoir box, each with enough lead tubing attached to account for three if not four pumps.[39]

Three lengths of the tubing are in the museum's collection. One is ca. 4.6 m (15 ft) long with a reservoir box and piston tube at one

38 Ships' Bilge Pumps

Fig. 22. (left) The lead pump from the San José *(1733), resting upside down. The dale extends to the left from the reservoir box. A bronze gravity valve is mounted in the end of the piston tube opposite the box. (Courtesy Florida Division of Historical Resources)*

Fig. 23. (right) The San José's *pump reservoir and piston tube, standing upright. Note the date of manufacture cast in relief at the top of the box. (Courtesy Florida Division of Historical Resources)*

Fig. 24. The lower extremity of the San José's *piston tube, with part of the bronze gravity valve visible. (Courtesy Florida Division of Historical Resources)*

extremity. The second exhibits a lower fitting that is an expanded section of tubing, and the third is a segment that originally joined the second, according to museum staff. The combined length of the second and third sections is approximately 8.54 m (28 ft)—nearly the maximum workable height of a suction pump. Therefore, the upper-valve cylinder would have had to be positioned within the next foot or so. It is possible, of course, that these two segments did not fit together and that three different pumps are represented. Another shipwreck at Yorktown, designated 44-YO-88, was excavated by the Commonwealth of Virginia between 1982 and 1990. The vessel was preserved up to the main wale on the starboard side. At the bottom of the pump well, which measured 30.5 x 51.3 cm (12 x 20 in), was a portion of a lead pump tube ca. 2.133 m (7 ft) in length. It was ca. 10 cm (4 in) in diameter and had a wall thickness of about 1.27 cm (0.5 in).[40]

A second pair of pumps, RS 6 and RS 7 (Fig. 26), was purchased in Saint Thomas, U.S. Virgin Islands, by agents of the museum because the tubes are similar to, but smaller than, those recovered from the York River. The upper valve cylinders of this smaller pair are

Fig. 25. Lead pump components from the York River, now held in the Mariners' Museum, Newport News, Virginia. Above: a reservoir box with dale outlet joined to the upper portion of the piston cylinder (RS 2). Below: the expanded lower extremity of tube RS 3. (Photos by author, courtesy the Mariners' Museum, Newport News, Virginia)

made of copper. No seams are evident on the lead tubing, nor is anything known of their history. Similarly, a copper upper-valve cylinder retrieved from the China trader *Rapid* (1811), an American East Indiaman that wrecked near Point Cloates, Australia, still has a lead nozzle attached to its bottom (Fig. 27).[41] A large-diameter (18.8 cm = 7 in) section of lead tubing was salvaged from the CSS *Florida*,[42] which sank in Hampton Roads, Virginia, in 1864. Its identification as part of a pump is not positive, but the large diameter suggests as much.

Different ways of manufacturing lead tubes were gradually developed. As noted earlier, the simplest method was to roll a narrow sheet of lead into cylindrical form and solder the edges together. A second technique, developed in the eighteenth century, entailed casting a section of tube in a mold, drawing it out until only a few inches remained in the mold, and pouring another section that thus joined it. This procedure could be repeated to produce tubes of consistent diameter and thickness in any desired length.[43] Tubes made in this way should exhibit mold marks opposite one another and at intervals along the length. A third method, that of drawing lead tubes from the molten state, was detailed in a U.S. patent granted to Burroughs Titus in 1801.[44] This suggests that drawn lead pipes may have appeared in the very late eighteenth century. Drawn pipes should display characteristic striations down the length.

Through the eighteenth and nineteenth centuries, as the ability

Fig. 26. A lead and copper pump (RS 6) recovered near Saint Thomas, U.S. Virgin Islands, also in the Mariners' Museum. The reservoir box, dale spout, and lower tube are of lead, but the piston cylinder is copper. (Photo by author, courtesy the Mariners' Museum, Newport News, Virginia)

Common or "Suction" Pumps   41

Fig. 27. Copper pump tube with lead nozzle from the American China trader Rapid (1811). (Photo by Patrick Baker, Western Australian Maritime Museum)

to work copper and bronze improved, these metals, although expensive, were used more and more in ships' pumps. In the eighteenth century, while other European navies were utilizing the chain pump aboard their warships, the French depended exclusively on the suction pump in their naval vessels. These French suction pumps were called royal pumps because they were of a distinct type used on the king's warships. A royal pump tube consisted of two wooden tubes with a cast bronze section set between them. The upper tube was 4.59 m long with a bore diameter of 16.3 cm, the lower one 4.98 m long and 12.2 cm in bore diameter. Five iron straps were affixed around each wooden section to prevent splitting and cracking. The royal pump piston worked within the bronze section, known as the *corps de fonte* (cast body). The wooden tubes were attached with four lag screws to a flange at each end of the cast body, the length of which varied with the shipyard for which it was made. At both Brest and Toulon the length was 97.5 cm, at Rochefort 86.7 cm.[45]

Examples from shipwrecks on opposite sides of the Atlantic show that there was indeed variation in the lengths of *corps des fontes*. One

Fig. 28. Corps de fonte *of a French royal pump recovered from the Fort Louisbourg wreck (mid-eighteenth century). A bronze piston valve worked within this cast-bronze cylinder, which was fastened between two wooden tubes. (Courtesy Canadian Parks Service)*

Common or "Suction" Pumps 43

Fig. 29. The bronze upper-piston valve of a French royal pump recovered from the Fort Louisbourg wreck (mid-eighteenth century). (Photo by author, courtesy Canadian Parks Service)

Fig. 30. Details of the bronze upper-piston valve from the Fort Louisbourg wreck. (Photos courtesy Canadian Parks Service)

from a mid-eighteenth-century wreck at Fort Louisbourg, Nova Scotia, is 95.5 cm long (Fig. 28). Another, from the wreck of *L'Impatiente* (1796) found off the coast of Ireland, measures 84 cm.[46] Their lengths may well indicate Brest or Toulon as the port of origin of the Fort Louisbourg vessel, and Rochefort for that of *L'Impatiente* (table 1).[47]

An upper valve made of bronze and weighing over 9 kg (nearly 20 lb) was found with the cast body from the Fort Louisbourg wreck (Figs. 29, 30). It closely resembles one in a Danish plan of a French suction pump,[48] except that a metal valve guide positioned at the bottom of the spear in the plan is missing from the archaeological example. The Danish drawing also illustrates a wooden lower valve with a metal stirrup of the basal ring type.[49]

TABLE 1. DIMENSIONS OF FRENCH *CORPS DES FONTES*

	Length (cm)	Inner Diameter (cm)	Wall Thickness (cm)	Flange Diameter (cm)	Weight (kg)
Brest and Toulon (ca. 1780)	97.5	16.3	1.4	33.9	146.7
Fort Louisbourg (ca. 1750)	95.5	16.5	2.4	35	>113
Rochefort (ca. 1780)	86.7	16.3	1.4	33.9	146.7
L'Impatiente (1796)	84	17	—	38	125

The undated lead and copper pumps (RS 6 and RS 7) in the Mariners' Museum are probably early examples of the use of a lead reservoir with a copper upper-valve cylinder. This conclusion can be drawn by comparing them to the lead pump with bronze valves from the *San José* (1733), the late eighteenth-century Yorktown pumps (RS 2 and RS 3), and the copper and lead valve cylinder tube from the *Rapid* (1811). RS 6 and RS 7 were most likely manufactured in the very last years of the eighteenth century and probably denote a transitional phase in the utilization of the two metals. The section of a bronze valve chamber from HMS *Pandora* (1791)[50] and the *Rapid*'s copper and lead tube demonstrate that by the late eighteenth century, both warships and merchant ships were beginning to use pump tubes of bronze and copper.

There were other advantages to the use of bronze and copper, as seen on the Civil War period vessels USS *Cumberland* and USS *Hartford*, whose tubes were copper, valves bronze,[51] and spear and brakes iron. The bore diameter of the *Cumberland*'s bronze cylinder is 22.1 cm, the inner diameter of the *Pandora*'s cylinder is 17 cm.[52] Wooden valves usually measure 9–12 cm in diameter. So, given the same piston stroke, a larger tube diameter would obviously yield a greater output. With this improved efficiency, bronze and copper pumps were replacing chain pumps on smaller warships in the first half of the nineteenth century.

At the beginning of the eighteenth century, iron pump tubes were finding important applications but were small in diameter; wooden pump tubes were still in general use. The only tools available for boring iron tubes were those used to make cannon.[53] The introduction in 1712 of the Newcomen steam engine, the first practical steam engine, therefore posed new and difficult manufacturing problems for ironmasters. A steam engine required that the piston and cylinder fit together within very close tolerances. Initially these cylinders, ca. 76–152 cm (30–60 in) in diameter, were ground and filed at considerable cost.[54] Further developments in the steam engine could occur only as the tools and processes for finishing the precision-fit parts were improved and as materials like iron were enhanced in quality and strength and became more readily available.[55] The first notable advancement in boring machines, patented by John Wilkinson on January 27, 1774, permitted the boring of cannon from solid castings. By 1776, Wilkinson was producing cylinders with a higher degree of precision than had ever been achieved.[56] Even so, it is difficult to determine exactly how the development of these boring machines affected metal pump fabrication. The care required for steam engine cylinder and piston production was not necessary for water pumps, especially those operating on ships. Nevertheless, the latter probably did benefit from such advances, as might be seen in the *De Braak* (1798) and *Pandora* (1791) pumps.

Numerous variations in the design of the common pump are docu-

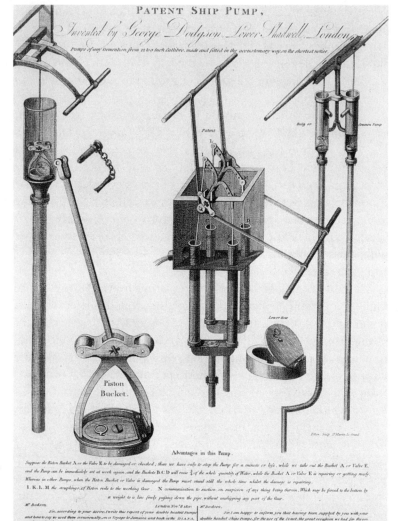

Fig. 31. Dodgeson's pump of 1799. Note the arrangement of the piston cylinders and the brake. (Courtesy National Maritime Museum, Greenwich)

mented in patent records and other sources, but not all were successful nor were all successful designs adapted to shipboard use.[57] One that was effective and used on ships was Dodgeson's pump, invented in 1799 (Fig. 31). Two upper-valve cylinders were connected by a T to one tube that extended to the bottom of the vessel. With a single piston in each upper cylinder, the use of a double-action brake generated a continuous flow up the lower tube. Two of these as-

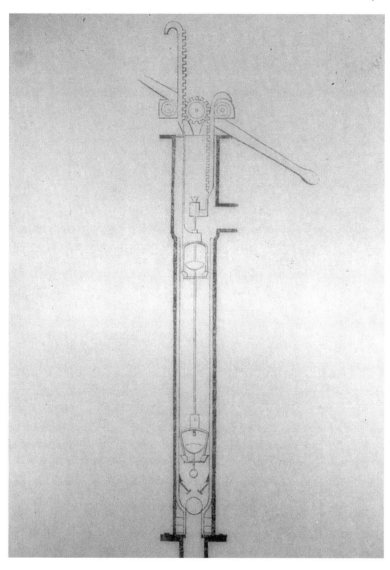

Fig. 32. Taylor's pump (1780), in which two pistons worked, one rising while the other fell. (Ewbank 1842, 226 fig. 91)

semblies were mounted together, providing a total of four cylinders.

Another version was the multiple-piston pump, in which more than one piston valve was in one cylinder. These pumps were popular on ships in the late eighteenth and early nineteenth centuries. In Taylor's pump (1780), each of two pistons was attached to its own rod so that one piston rose as the other fell. The rod for the lower piston acted as the guide for the valve of the upper one (Fig. 32). The top of each rod terminated in a rack that was moved by a cog

Fig. 33. One way a double-piston pump could be operated. A drum was turned back and forth to drive the piston rods. (Ewbank 1842, 227 fig. 92)

Fig. 34. A double-piston valve from the wreck of HMS Tribune (1796). The upper valve (left) was affixed to a rod (not shown) such that it could slide over the rod fastened to the lower valve (right). (Courtesy Canadian Parks Service)

wheel.[58] The wheel could be driven by a double-action quadrant brake or by a drum that was turned back and forth by a rope wrapped around it and pulled by a team of men in "tug-of-war" fashion[59] (Fig. 33). As a result, the pump emitted a more continuous flow of water, moving twice the volume of a single-piston common pump with the same bore and stroke.

The multiple-piston pump was used by the British navy from the 1790s through the early years of the nineteenth century, and by the French and Dutch most likely in the first half of the nineteenth century and later.[60] The pistons and rod of just such a pump were recovered by Canadian Parks Service personnel from the wreck of HMS *Tribune* (1797), a fifth-rate of thirty-six guns (Figs. 34, 35). The rod passes through the center of the upper piston and is attached to the staple of the lower piston. A second rod would have fit into a square socket in the stirrup of the upper-piston valve. The valves and rod are of bronze; the tube in which they operated was probably of the same material. It should be noted, however, that the *Tribune* was cap-

Common or "Suction" Pumps 49

Fig. 35. Details of the double-piston valve from the wreck of the Tribune. Top: the upper valve; bottom: the lower valve. (Courtesy Canadian Parks Service)

tured from the French by the British in 1796 and incorporated into the British navy. A year later she was wrecked on Thrum Shoals, Halifax.[61] Although the ship was probably refitted by the English, it is uncertain whether the pump is English or French, for no markings were found on it.

A double-piston pump was found on the wreck of HMS *De Braak*, which sank off the Delaware coast in 1798 (Fig. 36). She was built as a cutter in Holland for the Admiralty of the Maze in 1781. Captured by the British in 1797, she was extensively refitted as a brig and incorporated into the Royal Navy.[62] Two well-made bronze and cop-

Fig. 36. Elements of the double-piston pump recovered from the wreck of HMS De Braak (1798). (Courtesy Delaware State Museums)

per pump assemblies were recovered from the wreckage, while the lowermost sections, of wood, were left in situ on the wreck. The bronze tubes are 1 m long with an inner diameter of 18.4 cm and tapering lower ends. A groove spirals along the length of each as though threads had been cut into them, indicating they were probably turned on a lathe. The bronze lower valves of the pump were threaded and may have screwed into these tapered lower ends, based on the *Pandora*'s example. The tapered part of the latter's bronze tube was a separate piece that screwed onto the main body.[63]

The piston valve assemblies (Fig. 37) consisted of a lower-piston rod and the upper- and lower-piston bodies (each with a basal ring probably of brass but perhaps bronze), a valve claque weight that was attached to the valve leather, and a valve claque "keeper" that held the leather in place. The lower-piston rod passed down through a hole in the upper-piston valve body, and a nut tightened at the threaded lower end of the rod held the lower-piston valve in place. To prevent water from passing between the piston and the tube wall, a leather gasket wrapped around each piston body was held in place

Common or "Suction" Pumps 51

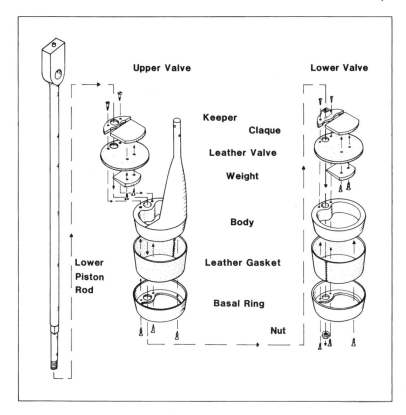

Fig. 37. Schematic of the double-piston valve assembly from HMS De Braak's pump.

by a basal ring. The basal ring, valve claque keeper, and valve claque weight were all secured with handmade screws. Each screw fit only its own hole and was marked, as was its position, with different arrangements of punched dots.[64]

In 1825, three pistons were mounted in one cylinder by Jonathan Daunton of London. The three piston rods worked one within another, telescope fashion, and were probably connected to a camshaft operated by a winch or crank. This pump type was also used in the British navy for many years. Another modification was made a little later by Stone and Depthford, who placed four pistons—two attached to each of two rods—in one cylinder.[65] This pump had the ability to continue working if one of the valves became clogged or needed repair. A modified version of the Stone-Depthford pump that provided easier access to the valves was patented by Blundell and Holmes in 1876,[66] indicating that multiple-piston pumps were still being used in the second half of the nineteenth century, at least on land.

It was Ewbank's opinion in 1842, however, that this type of pump

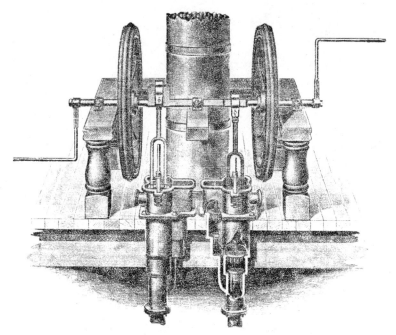

Fig. 38. This illustration shows the typical arrangement of pumps on large sailing vessels from the mid-nineteenth century into the twentieth. (Hyde Windlass Co. catalog 1902, 150)

was essentially a double pump in one cylinder requiring twice the length of cylinder and twice the work to operate it. He thought that two separate pumps were preferable to one double-piston pump because they were less complex, less likely to malfunction, and had a longer stroke. Also, if one became fouled or broke down, the other could continue. Based on these considerations, Ewbank believed the British navy abandoned the double-piston pumps.[67]

Around the 1850s, the typical arrangement of pumps on packets and clipper ships consisted of a pair of pumps placed just aft of the mainmast (Fig. 38). The general configuration comprised two piston rods connected to a camshaft that was mounted on the main fife rail. On one or both ends of the shaft, inside the fife rail, was a large iron flywheel that helped maintain the momentum of rotation. At each extremity of the camshaft were removable cranks by which the crew operated the pumps.[68] Through the last half of the nineteenth century and the first decades of the twentieth, the pumps of most medium and large merchant vessels were of this type. The mechanisms were by then made completely of iron, copper, or bronze parts, excluding leather or rubber gaskets.

Common pumps on warships of the eighteenth and nineteenth centuries had secondary functions: providing water for washing and

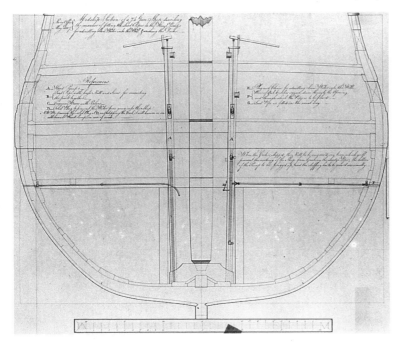

Fig. 39. An Admiralty plan dated 1807 for a seventy-four-gun ship shows how seawater was fed into the well or the pumps through lead pipes for cleansing the ship. Stopcocks at the sides of the ship controlled the flow of water. A plug at the bottom of the right pump kept the water from entering the bilge. (Courtesy National Maritime Museum, Greenwich)

fire fighting. Because the bilge water was usually foul, fresh seawater was brought into the hull to flush out stagnant bilge water or it was brought directly to the pumps. According to Arthur Bugler, the two common pumps on HMS *Victory* (built 1765, rebuilt 1801) pierced the bottom of the hull and provided only clean salt water.[69] Bugler does not seem certain of this statement, however. Such an arrangement would have meant fashioning a hole through more than 30 cm of oak frame and planking and piercing the copper sheathing without allowing leakage. The water pressure against the hull at this point would have been considerable, and leaks would have been very difficult to avoid.

A British plan dated 1807 (Fig. 39) illustrates another method of furnishing the pumps with seawater. A lead pipe pierces the side of the ship below the waterline and either empties into the well or feeds directly into the pump tube. The flow of seawater through each pipe is controlled by a stopcock near the side of the ship.[70] The tube of the pump on the right in this schematic is fitted with a removable plug near its bottom. With the plug removed and the stopcock shut, the pump could be used in the usual manner. Another diagram, dated 1825, shows a wooden pump set into a cistern consisting of a lead-lined wooden box resting on brackets at the side of the well, up off

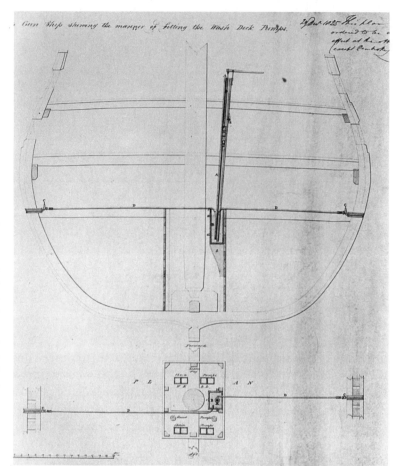

Fig. 40. An 1825 plan of a wash pump set into a cistern mounted on brackets in the well. Seawater is brought to the cistern by lead pipes. A note states that this system was adopted for all shipyards except Pembroke. (Courtesy National Maritime Museum, Greenwich)

the bottom of the ship. As in the arrangement shown in the 1807 drawing, the cistern was filled with seawater by means of lead pipes piercing the sides of the ship and regulated by stopcocks (Fig. 40).

Frank Howard states that common pumps had an advantage over chain pumps in that they could put water under some pressure, which could then be used for fighting fires.[71] The only pressure that a common pump can develop, however, is that of the head (i.e., the pressure created by the weight of a water column). Only a force pump can put water under any significant pressure. Therefore, it is submitted here that force pumps were used in fire-fighting engines, which were sometimes placed on larger ships,[72] and that common pumps probably supplied the force pumps with seawater.

The common pump first appeared in Italy in the beginning of

the fifteenth century and by the sixteenth century had come into general use. In shape, the wooden upper and lower valves varied only slightly throughout the study period. The use of metals began in the early sixteenth century with lead sieves and the Molasses Reef wreck lead upper valve. Copper and bronze came into general use in the eighteenth century, and iron became prevalent in the late nineteenth century. The multiple-piston pump, used in the late eighteenth and early nineteenth centuries, was the most successful variation of the common pump. In conjunction with stopcocks, common pumps were also used as wash pumps and to supply water for fighting fires.

5

The Chain Pump

As described by William Falconer, the chain pump (Fig. 41) was "a long chain equipped with a sufficient number of valves, at proper distances, which passes downward through a wooden tube, and returns upward in the same manner on the other side. It is managed by a *roller* or *winch*, whereon several men may be employed at once, and thus it discharges, in a limited time, a much greater quantity of water than the common pump, and that with less fatigue and inconvenience to the labourers."[1] Additional parts included the winch handle, by which the drive wheel was turned; the cistern; and a roller or idler drum at the bottom of the round chamber (ascent tube) that helped guide the chain into the tube.

The origin of this machine has been attributed to the Chinese, who used it both in agriculture for irrigation and in their ships for pumping the bilges.[2] Gaspar da Cruz, a Portuguese Dominican priest, gave this description in 1556:

> The Chinese do use in all things more sleight than force. Even in large leaky ships, one man seated alone, moving his feet as in climbing a staircase, pumps it out in a short time. Pumps [are] made of many pieces in the manner of water wheels laid alongside the side of the ship, between rib and rib. Each piece of wood half a yard more or less, one quarter [of it] well wrought. In the midst of this piece of wood is a square little board almost a hands breadth and they join one piece to another in such a manner as it may double [bend?] well. The joints, all close, are within the breadth of the little boards of every one of the pieces, for they are all equal:

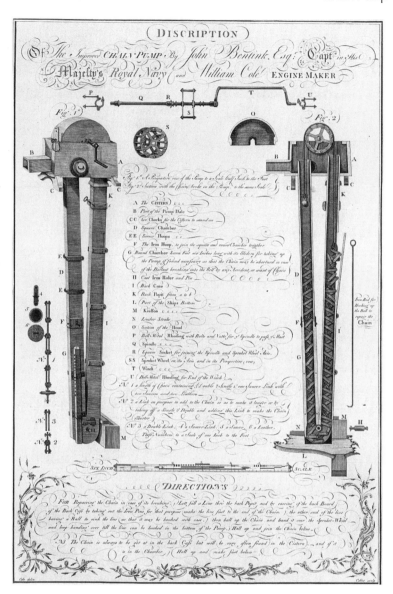

Fig. 41. The Cole-Bentinck chain pump was developed in the latter half of the eighteenth century. (Courtesy National Maritime Museum, Greenwich)

and this manner of pump bringeth up so much water as may be contained between the two little boards.[3]

Juan Gonzales de Mendoza, a Spaniard, wrote in 1585: "The pumps which they have in their ships are much differing from ours, and are far better. They make them of many pieces, with a wheel to draw water, which wheel is set on the ship's sides within, wherewith they do easily cleanse their ships, for that one man alone going at the wheel doth in a quarter of an hour cleanse a great ship although she leak very much."[4]

The earliest mention of the use of chain pumps on European ships is made by Sir Walter Raleigh, who lists the chain pump, bonnets, stun'sails, and the anchor capstan as improvements introduced into the English navy during the last half of the sixteenth century.[5]

Chain pumps were, however, known in ancient Europe and were used on ships. The remains of machines made of rope and wood and possibly leather have been recovered from shipwrecks of the Roman and early Byzantine periods.[6] Sometime during the Middle Ages, the chain pump seems to have disappeared in Europe because it was reintroduced in Italy early in the fifteenth century. The Italian engineer Mariano Taccola recorded in his notebooks dated to 1431 that of the various pumps he showed, the chain pump was something entirely new and different. He attributed it to the Tartar peoples of eastern Europe.[7] From there it is believed to have passed via Italian merchants to the rest of Europe. More extensive European contact with the Far East, beginning in the sixteenth century, permitted more medieval Europeans to experience different and advanced cultures for the first time. Ideas traveled in both directions, and the concept of chain pumps on ships could well have been brought back by early explorers.

In fact, by Raleigh's time, Europeans were already utilizing a perfectly effective type of chain pump in their mines and may have simply adapted it for use on ships. Georg Agricola[8] illustrates a *paternoster* or "rosary" pump, a descendant of the Tartar pump.[9] The *paternoster* pump was a crude form of chain pump with valves made of leather bags stuffed with hair or cloth and fastened to a rope or chain. Sixteenth-century shipboard chain pumps are described as having blocks of wood as valves, to which rags were attached.[10] This description more readily evokes the European chain pump than the Chinese machine.

We have no detailed descriptions of chain pumps of the late sixteenth century or of the seventeenth, and dictionaries of the time do not provide very specific definitions. N. Boteler and Sir Henry Manwayring describe a chain pump as a chain full of burrs that goes around a wheel.[11] It was said to be the best type of pump and to have discharged the most water. Manwayring adds that it was easy to repair because spare "esses" were always carried on board. The word *ess* refers to the shape of the chain links—that of an S—although standard round or oval interconnecting links were also used.[12]

Burrs were valves or disks that were attached at intervals to the chain. The term *burr* returns us to the burr pump, suggesting that the valves may have been wooden, as stated by Frank Howard.[13] By the late seventeenth century, they were made of metal, as described thus: "Pump chain burrs are round thin pieces of iron, very little less than the bore of the pump, which are placed between every length of chain and on each of them the leather is put for bringing up the water."[14]

The drive wheel consisted of a solid wooden wheel with "sprockets" positioned around the circumference. Each sprocket was horseshoe-shaped with a tang or spike that projected from the bottom of the curved end and was driven into the wheel.[15] The crankshaft was fitted through the center of the wheel. An English chain pump for a fifty-gun ship, known from a Dutch drawing dated 1736,[16] is very similar to the pump described by Thomas Blanckley. Although the drive wheel appears to have been made of iron, the side view illustrates the same type of horseshoe-shaped sprockets. The chain comprises round interconnecting links instead of S-shaped links. Both the round- and the S-link chains presented problems beyond those of maintenance (i.e., having to remove the chain from the pump to replace the leathers). The chains tended either to jerk back or slip off the drive wheel or to bind and break under the weight of the water.

In the eighteenth century, the British navy reviewed numerous "improved" chain pumps. None was found to be satisfactory until William Cole and John Bentinck introduced theirs in 1768 (Fig. 41).[17] Their design of the chain and the drive wheel decreased link wear and also reduced the probability that the chain would slip or jerk back under the load. The chain consisted of alternating single- and double-link bars connected by pins secured with cotter keys. The drive wheel was now two cast-iron wheels connected by bars around the circumference that caught nibs or hooks on the links. With fewer men working it, the pump discharged more water than did other chain pumps. In a trial on HMS *Seaford*, the new pump with four men at the crank pumped one ton of water in 43.5 seconds. In contrast, the old type of chain pump with the same number of men required 83 seconds to move the same amount of water.[18]

One of the most important features of the Cole-Bentinck type of chain pump was the ease and speed with which it could be disassembled and repaired. The chain links were cast the same size and were therefore interchangeable; link pins could be made to a standard length because the widths of the links were consistent; uniform cotter keys could be cut easily from a sheet of copper; the valve cylinders or disks could be prepared ahead of time and kept ready; wooden parts could be fabricated by the ship's carpenter; and lastly, all these stores could be kept on board.

A variety of tools was used to effect repairs, including tongs, hooked poles, and counterweights.[19] Instructions for remounting the chain and maintenance of the pump are given by Falconer.

> First, try if it [the chain] can be hooked in the pump; if it cannot, turn the chain gently back until it will go no farther; then you may be sure the end of it is down at the bottom of the pump. Ease off the back-case, by taking out the iron pins, haul out the chain, and put it aside; apply a hook-rod with the hooks toward you, and hook the end of the chain at the bottom of the pump: when you have it fast, let the men above put back the chain gently, to enable you to haul it up to the wheel, where the broken link is to be taken out and a new one put in. Observe always to put the stops or hooks of the new link in the same direction as those on the chain you join them to, overhaul the pump, examine the forelocks, make all safe, put on the back-case, and the pump is ready to work.
>
> It is strongly recommended to all persons who may have charge of a chain pump, to practice the method of recovering the end of the chain until they are expert in the performance of that operation, and putting in a new link.
>
> In case the chain draws to one side of the wheel in working, raise the rhoding [bushing] on that side until you find it keeps an equal distance from the sides. To prove what water the pump loses in working, turn the winches with a velocity that will keep the water to the top of the pump, but without flowing over; observe how many revolutions per minute are required to do this, for so many are lost in that time.[20]

The ironwork was to be covered with tar or varnish or, if those materials were not available, with tallow. By protecting the iron from corrosion in this way, its strength was preserved for a longer period.[21]

The Cole-Bentinck pump was put on trial in British warships in 1770, but because the chain demonstrated a tendency to break, installations were halted in 1773. Design problems were rectified in 1774, at which time the Navy Board ordered its use on all warships.[22] Early in the nineteenth century, the chain was improved further by a Mr. William Collins, who replaced the cast-iron links with brass ones.[23] Thereafter, the chain pump was used by the British navy through the first half of the nineteenth century. Most northern European countries were using the chain pump in their warships by the end of the eighteenth century.[24] The French were a notable exception, even though they were employing it on land to empty cofferdams and dry docks.[25]

As early as 1776, the chain pump was found in warships of the Continental Navy,[26] but it fell out of favor in the United States Navy in the first decades of the nineteenth century. As a result, the British sloop of war *Cyane* had her chain pumps removed and replaced with common pumps after her capture and induction into the U.S. Navy in 1815.[27]

The reintroduction of the chain pump on larger U.S. naval warships occurred in the late 1830s. A description of the pump placed in service aboard the USS *Independence* in 1837 is provided by Thomas Ewbank;[28] two similar pumps were used aboard the USS *Ohio*. The chain consisted of copper links designed like those of the Cole-Bentinck pump. The two-foot-diameter cast-iron drive wheel, with twelve spokes radiating from the hub, was rimless. A notch or fork at the end of each spoke caught the chain as the wheel revolved. The tops of the pump tubes were exposed at the upper deck, the cistern having been eliminated. The back case (descent tube) stood ca. 25.4–30.5 cm (10–12 in) above the deck to prevent water from returning to the hold. Both the back case and the round chamber were made of copper and measured ca. 6.7 m (22 ft) in length and ca. 17.8 cm (7 in) in diameter. The unique features of this model were the comparatively long lengths of the copper tubes and the arrangement of the drive wheel. The latter was similar to that of early chain pumps, in that a fork on the drive wheel engaged the chain. Here, an old idea was reborn with the aid of better engineering and technology.

The chain pump was a great improvement over the common pump. The former could move a greater quantity of water faster and was easier to work than the latter. In the seventeenth century, a man would work at the common pump until he had made 200 strokes (ca.

6–7 min), whereas a spell at the chain pump was reckoned in "glasses" (probably one half-hour glass).[29] In the late eighteenth century, a spell at the common pump lasted about five minutes, while thirty minutes was still the standard interval at the crank of a chain pump.[30]

Yet the chain pump did have its disadvantages. The many leather disks of the valves had to be replaced about every twenty days;[31] the efficiency of the pump itself was only about fifty percent;[32] and many men were required to turn the crank handle. For these reasons, the chain pump was not used on smaller warships and never gained favor on merchant ships.

The only intact chain pump known to this author is on board HMS *Victory* (built 1765, rebuilt 1801).[33] The valves on its chain are

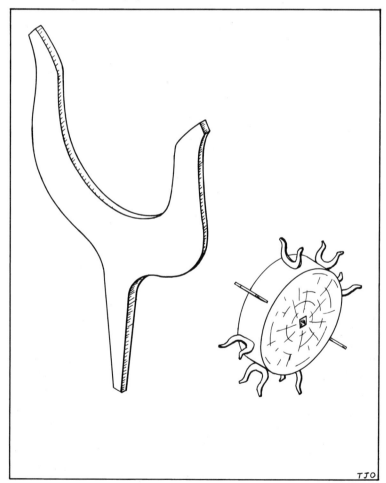

Fig. 42. Sprocket and drive wheel of an early type of chain pump, as reconstructed from the iron sprockets found on the Santo António de Tanna (1697).

ca. 91 cm (3 ft) apart, the round chamber is made up of short iron sections bolted together, and the lower structure around the roller is also metal. The back case consists of boards.[34] Various chain pump parts have been found, however, on the wrecks of the *Santo António de Tanna* (built 1681, sunk 1697); HMS *Charon* (built 1778, sunk 1781); HMS *Pandora* (refitted 1789, sunk 1791); and HMS *Pomone* (built 1805, sunk 1811). A few objects found from another site in the York River in Virginia, possibly the final resting place of HMS *Fowey* (1781), might also be chain pump remains.[35]

Various pump elements recovered from the *Santo António de Tanna*, a Portuguese warship sunk in Mombasa Harbor, Kenya, in 1697, represent the only certain example of an early type of chain pump found to date.[36] They closely resemble the descriptions given by Boteler, Manwayring, and Blanckley. The remains consist of lengths of the chain, with parts of the metal and leather valves in place, and a few horseshoe-shaped sprockets. Both ends of the sprockets are flared outward (Fig. 42), but whether the sprockets caught the valves themselves or just the chain is not known. All of these components were found in the remnants of a toolbox, not loose in the pump well, so they most likely represent spare parts.

The chain links are S-shaped with their eyes oriented ninety degrees to each other (Fig. 43). Special links held together the metal disks and leathers of the valves, which were spaced about 50 cm apart. A "stirrup" that had an expanded bottom bar with a hole in its center formed the top element of the link. The bottom element consisted of an eyebolt-shaped piece whose straight shaft received first a metal disk and then several leather ones. The shaft was then inserted through the hole in the base of the stirrup. The end of the straight shaft was then flattened to secure the entire assembly.

This type of valve construction was labor-intensive, and disassembly for leather disk replacement must have required a great deal of hammering. For this reason, it was recommended that a second chain always be kept for each pump. While one was in use, the links of the other could be repaired or the leathers replaced. Two different types of tools could be used to open the links during maintenance and repair operations. One was a fid or marline spike—common maritime tools. The other was a "bolster," which was "a round piece of iron with a hole in the middle, and was for opening an Ess or hook when any want shifting."[37]

Two holes within the pump well of the *Santo António*, for the

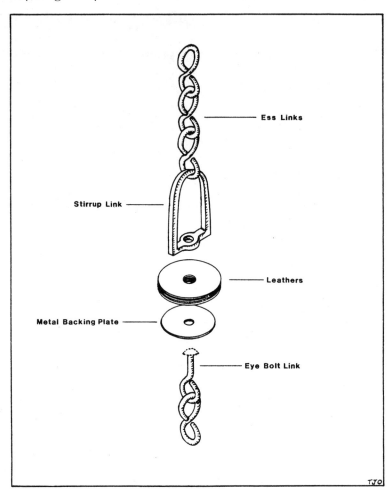

Fig. 43. Chain and valve assembly of an early type of chain pump, as reconstructed from elements found on the Santo António de Tanna (1697).

seating of its suction pumps, were the only openings in the ceiling planking. No evidence for the location of the chain pump seatings was found anywhere in the bottom. In all probability there were two chain pumps, one set to each side of the keelson in the manner of the common pumps.[38] One round chamber (ascent tube), a hollowed log with remnants of the chain still in it, was found at the forward end of the pump well. Other parts include two fragments of planking joined at their edges to form a ninety-degree angle that represented part of the back case (descent tube). A small iron hinge plate may be from a door in the back case, and an iron plate with a hole in the center possibly acted as a bearing for the lower roller. Based on the physical evidence as well as descriptions and illustrations of early chain pumps, I have reconstructed the *Santo António*'s chain pump (Fig. 44). The chain pump fragments from this wreck are the oldest physical remains of a post-Byzantine shipboard chain pump.

Two wrought-iron objects thought to be parts of a chain pump were recovered from one of the Revolutionary War wrecks, prob-

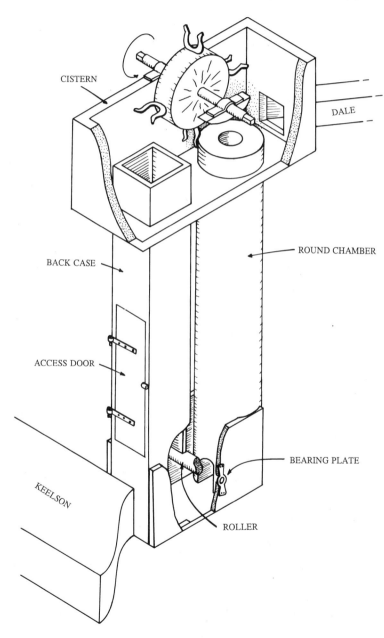

Fig. 44. Cutaway view of the Santo António de Tanna's reconstructed chain pump (chain omitted for clarity). (Drawing by author and Maria Jacobsen)

ably HMS *Fowey* (1781), in Virginia's York River in the 1930s. One object comprised two four-spoked, rimless wheels. The four spoke extremities of one wheel were bent, threaded, and bolted through holes at the spoke extremities of the other wheel. The second item looks like a bar link with a hole at each end.[39] The identification of these objects as chain pump parts is open to question, but they could represent a transition between the older variety of chain pump and the Cole-Bentinck model. The pieces are not part of HMS *Charon*'s pump (see below), and the only other warship known to be at the bottom of the river is HMS *Fowey*, a twenty-four-gun sixth-rate built in 1749.[40] But because the *Fowey* was an old ship and had been on station in the colonies for many years, she was probably not equipped with the newer Cole-Bentinck type of chain pump. Archaeological work on the wrecks of the Cornwallis fleet by the state of Virginia has not produced any similar artifacts, or any items identifiable as chain pump parts, other than those from HMS *Charon*.

The *Charon* was General Cornwallis's supply fleet flagship and was sunk in the York River during the siege of Yorktown, Virginia, in 1781.[41] Excavations on this forty-four-gun, fifth-rate British warship rendered the remains of a Cole-Bentinck type chain pump. Her two chain pumps were enclosed within the pump well, along with the main mast step and the suction pumps (Fig. 45). Two shot lockers, one abaft

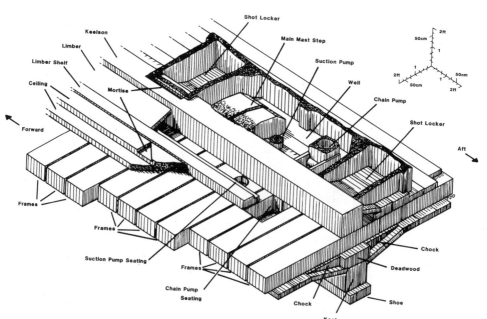

Fig. 45. View from the port quarter of the midships area of the Charon, *showing the remains of the shot lockers and the well, which contained the main mast step, suction pumps, and chain pumps.*

and one forward of the well, shared common walls with the well. This particular arrangement of shot lockers, mast step, and pumps was one of the features that enabled the excavators to identify the wreck as that of the *Charon*.[42]

The seating for each chain pump consisted of notches cut into the adjacent molded surfaces (vertical faces) of two frames. Planks placed at the fore and after sides of this opening supported the round chamber, the back case, and the cast-iron roller or idler drum that guided the chain into the round chamber (Fig. 46). These parts of the *Charon*'s chain pumps, which formed a "lower structure" that supported the rest of the pump, were securely set into the fabric of the hull. In contrast, Falconer's illustration of a chain pump shows the

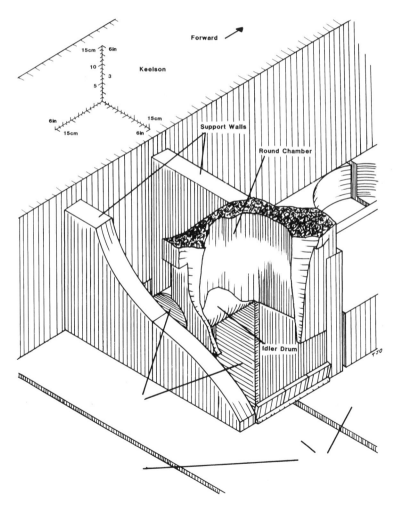

Fig. 46. The seating of HMS Charon's *chain pump. The idler drum was positioned between the vertical boards that supported the round chamber and back case (not shown).*

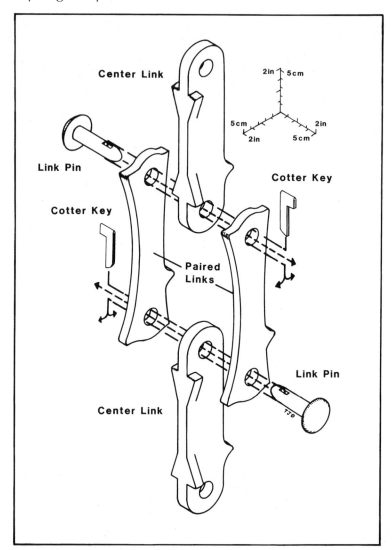

Fig. 47. Chain assembly of HMS Charon's chain pump. Note the nib on the inner edge of each link and the unusual triangular projections on the outer edge of each center link.

entire lower structure set above the frame tops and the bottom of the descent tube resting partly on the keelson.[43]

The bottom 30 cm (ca. 12 in) of the round chamber (the outboard tube) of the *Charon*'s starboard chain pump were preserved. Its preserved inner diameter is ca. 17.8 cm (7 in), and its base was cut to fit the boards of the supporting walls of the lower structure. Figure 41 shows that the round chamber was a bored tube that did not extend up to the cistern on the main deck. The intervening section was made of plank, so a horizontal line on the ascent tube just above the third waterline on the *Charon*'s plan must represent this feature.[44] No evidence for metal sleeves in any of the tubes was found. Remnants of the back case indicate that it was a box made of vertical planks and therefore similar to those of the *Victory*'s pumps.[45] Its sides were removable to make clearing and repairing the pump easier.

The basic design of the chain (Fig. 47) is the same as that illustrated in figure 41: single center links alternate with paired or double links that are joined at each end by what appear to be copper-coated iron pins. Cotter keys cut from a sheet of copper were inserted through slots in one end of each pin to secure the links. The shape of each key was that of an inverted L, created by folding one arm of a T-shaped piece of copper over the other.

The construction of the *Charon*'s chain, which is identical to that of the *Pandora*,[46] was such that each link had a distinct inner and outer edge. A nib projected from each link's inner edge near one end to engage the bars of the drive wheel and to prevent the chain from slipping, which is in keeping with the Cole-Bentinck design. On the pumps of the *Pomone* and the *Victory*, hooks were used instead of nibs.[47]

A significant feature of the *Charon*'s and *Pandora*'s chain links not mentioned in any contemporary source is the presence of opposing triangular extensions that project on both sides of the outer edge of the single links near each end as well as the corresponding projections on the ends at the outer edges of the paired links. The bases of the triangles face the extremities of each adjacent set of paired links and touch them when the chain is straight (Fig. 47). This clever design meant the chain could not bend inward as it came off of the drive wheel or scrape and damage the inside of the tube or the valve leathers. Further, because the chain could bend in only one direction, it could not collapse down the tube and was accordingly easier to recover if it broke.

Another unique characteristic of the pump recovered from the *Charon* is the valve design. Falconer describes chain pump valves as "two circular plates of iron with a piece of leather between them."[48] Such valves would be similar to those in the pumps of the *Victory*, the *Pandora*, the *Pomone*, and the *Santo António*.[49] The type of valve found on the *Charon*, however, is a short cylinder of unidentified cast metal with a central hole for the chain (Fig. 46). A projecting lip at the base of the cylinder supported a piece of leather that encased the cylinder and was fastened by a metal band (evidenced by stains). The reason for this design variation is unclear. Perhaps it was an attempt to reduce the wear on the leather and decrease the frequency of replacement. In any case, it does not appear to have been successful because this type of valve is not known from any other context. The spacing of valves on the chain could not be determined for the *Charon*'s pump.

According to figure 41, the Cole-Bentinck valve assembly rested upon the support ledge of a single chain link and was secured with a cotter key inserted through the link. The arrangement on the *Charon*'s chain was different. Each valve was mounted on a single center link that had a triangular extension at only one end. The end without the extension passed first through the hole in the disk on the side with the lip, and the disk rested on the projection on the link. Two small, flat bars were fastened to the top of the valve, covering the hole on each side of the link. This provided a surface upon which the cotter key inserted through the single chain link could bear.

Other artifacts that may be remains of the *Charon*'s pump were found in the well area. A flat piece of cast iron with a curved outer edge and a threaded hole through the thickness is perhaps a piece of the outer edge of a drive wheel. The remains of more copper-covered pins of different diameters and longer than those required for the link pins may have been used to secure the tube in place at its base or to hold the sides of the back case together.

Excavations on the remains of HMS *Pomone* (1805) at the Needles, Isle of Wight, yielded a length of pump chain that is nearly identical to that of the *Victory*.[50] These chains were of the same type as the *Charon*'s but differed notably in design details (cf. Figs. 47–49). The chain links on the *Pomone* and the *Victory* were thinner, and the nibs on the interior edges of the *Charon*'s single links had developed into a hooking appendage at one end of each link and a similar but less hooklike protrusion at the opposite end. Both of these features were designed to catch the bars of the drive wheel. There were no other features on the edges or ends of the links. The pins joining the chain links were rivets on the *Pomone* except for those in the links holding the valves, which were fitted with a pin and cotter key at one end.[51] The *Pomone*'s chain was bronze, which agrees with Falconer.[52] The drive wheels recovered from the *Pandora* are like those illustrated in the Cole-Bentinck plan (Fig. 41).[53] Lastly, the pump valves from the

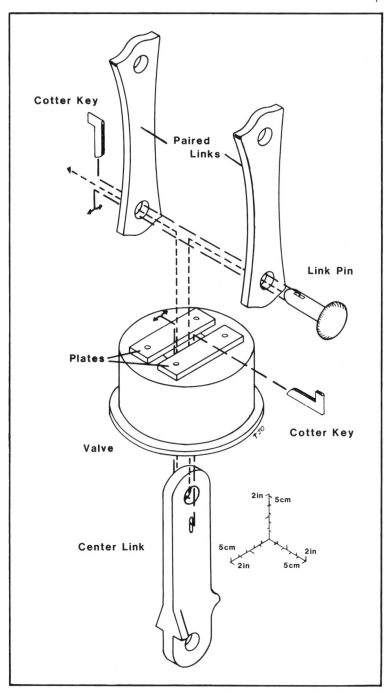

Fig. 48. The valve assembly of HMS Charon's chain pump, reconstructed here, represented a marked departure from the standard valve design.

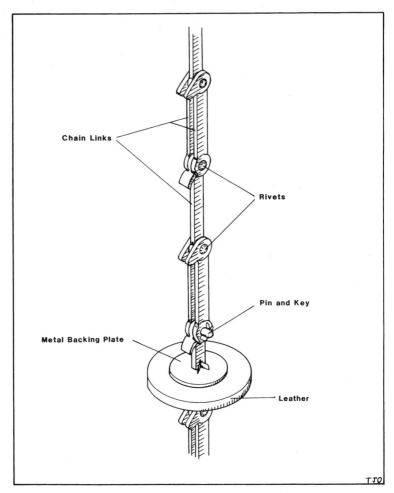

Fig. 49. A representation of the chain and valve assembly used on the Victory and the Pomone in the last years of the eighteenth and early nineteenth centuries.

Pandora, the *Pomone*, and the *Victory* resembled those described by Falconer: leather between metal disks.

The chain pump is an old machine that was used on Roman and Byzantine ships. Completely forgotten in the Middle Ages, the device was reintroduced from eastern Europe through Italy in the fifteenth century. The idea for using it on ships may have been brought from China. By the last quarter of the sixteenth century, the English placed it on their warships as a high-volume pump; from that time on it was used by most European navies except the French. The early version of the machine exemplified by the pump of the *Santo António de Tanna*, with its horseshoe-shaped sprockets and S-shaped chain links, was always subject to failure of the chain. Not until the late eighteenth century were chain pumps built with an adequate chain. The Cole-Bentinck pump, after some years of testing and development, became the standard pump for most large British warships. Even after this pump was accepted by the Admiralty, changes were made in the chain and valve assemblies (as shown by the *Charon*'s pump) in apparent experiments to improve it. The chain pumps of the *Victory* and *Pomone* were most likely the final form of the pump.

6

Later Pump Types

By the middle of the nineteenth century, the industrial manufacture of pumps had commenced. Iron became the predominant metal for all pump parts in the last half of this century, although copper and bronze continued to be used.

The suction pump was the type of bilge pump utilized most frequently in the nineteenth century, with quite advanced designs becoming available by around 1900. The Deluge or New Deluge suction pump (Fig. 50) is illustrated in catalogs issued in 1888 by Goulds and in 1890 by Smith and Winchester.[1] It was recommended for small vessels with hold depths of ca. 4.5–6.0 m (15–20 ft). In 1902, the Hyde Windlass Company offered the Hyde Deluge pump in three different sizes, to fit "the smallest coaster or fisherman up to schooners of twelve hundred tons register, having a vertical lift of twenty-four feet."[2] The cylinder diameters ranged from ca. 20–25 cm (8–10 in) and the pipe diameters from ca. 5–9 cm (2.5–3.5 in). The brakes of all Deluge pumps were bent near the end that was inserted into the pump so that when used with one side up, the working action of the brake was up and down; conversely, the action was back and forth when the other side was up. The pump was made so that the brake could be positioned on either side of the pump body and sometimes at the back as well.

The force pump and the diaphragm pump also appeared before the end of the nineteenth century. The force pump had been known to the Romans,[3] but significant technological improvements had increased the efficiency of this pump such that it had become truly use-

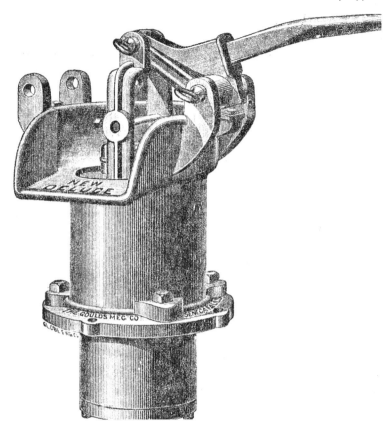

Fig. 50. New Deluge bilge pump. (Smith and Winchester 1890, p. 99 fig. 829)

ful. In this style of pump there is no valve in the piston. On the upstroke, water is pulled into the piston cylinder through an intake valve (the lower valve) and then forced out through an outflow valve (comparable to the upper valve of a common pump) on the downstroke. When the pump assembly included an air chamber, the water could be pumped out under pressure and used for washing or fire fighting. This pump was recommended for house, ship, and factory use (Fig. 51).[4] A two-cylinder force pump like this one is located in the Penobscot Marine Museum in Searsport, Maine (Fig. 52). Its cylinders, air chamber, piston rods, and other working parts were made of brass. Other force pumps were sold under the model names of Challenge and Alert (Fig. 53).

A type of diaphragm pump was developed that worked much like the common pump except that a flexible rubber diaphragm took the place of the upper piston. The outer edge of the circular diaphragm

76 Ships' Bilge Pumps

Fig. 51. (left) *Two-cylinder brass force pump. (Smith and Winchester 1890, p. 84 fig. 160)*

Fig. 52.(right) *A two-cylinder brass force pump at the Penobscot Marine Museum, Searsport, Maine. (Photo by the author; courtesy Penobscot Marine Museum)*

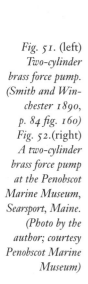
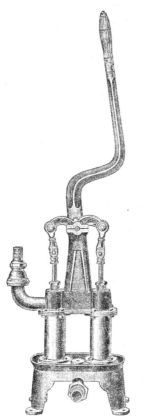
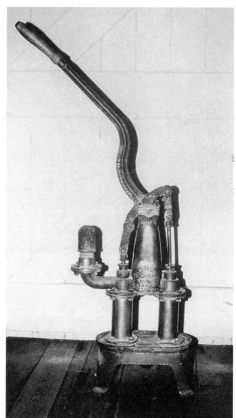

Fig. 53. (left and right) *Challenge and Alert double-acting force pumps. (Smith and Winchester 1890, pp. 85–86 figs. 562 and 747)*

Later Pump Types 77

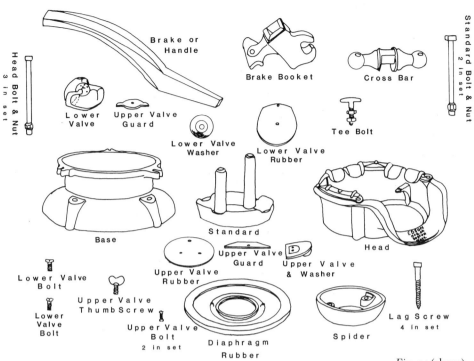

Fig. 54.(above) Parts for the Edson diaphragm pump. (Drawing by Helen DeWolf. After George B. Carpenter catalog 1903, p. 187)

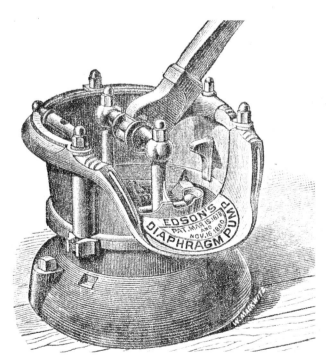

Fig. 55. The Edson diaphragm pump clearly showing patent dates. (Smith and Winchester 1890, p. 100 fig. 768)

was attached to the body of the pump. The upper valve claque was positioned in the center of the diaphragm and linked to the brake by a piece called the "standard" (Fig. 54). The Edson diaphragm pump carries patent dates of March 19, 1878, and November 16, 1880 (Fig. 55).[5] The Louds diaphragm pump, with a diaphragm ca. 32 cm (12.5 in) in diameter, could discharge ca. 5.7 l (1.5 U.S. gal) of water per stroke.[6] On all of these models, the brake could be operated from either side of the pump.

These types of pumps were the culmination of manual pump technology. As shipboard power increased in all sizes of ships and more efficient pumps were introduced, the crew could be relieved of the backbreaking work of pumping ship by running the pump with the ship's engine.

7

Summary and Conclusions

Because of the way they are built and the stress they undergo, all wooden ships leak to some extent—some more than others. To preserve a ship and keep the crew, the passengers, and the cargo safe, some method of expelling the bilge water from the hold is essential. An efficient bilge pump, therefore, is a required piece of equipment on any vessel, and an improved pump, like a better mousetrap, is always in demand. New technologies are accordingly reflected in the history of ship's pumps.

Wooden pump tubes throughout the period in question were made by boring holes through the centers of solid logs. The logs, carefully chosen for straight grain, were pierced by a small pilot hole, after which successively larger bits were used to widen the bore to the proper diameter. Wooden pump tubes were created in this way well into the twentieth century.

The burr pump was in routine service on ships in the early sixteenth century. In this period it consisted of the tube (which rested on a large, separate foot valve that formed the base) and the rod and burr. It had fallen out of widespread use by 1625, but because of its simple construction it never completely disappeared. A type of burr pump was discovered on one sixteenth-century wreck, and examples have been found on ships of the 1860s. Generally, the burr pump declined in popularity as the common pump's advantages gained increasing favor.

Following its emergence in Italy early in the fifteenth century, the common pump gradually became one of the more important hydraulic machines in Europe. When it was first employed in the bilges of ships is not known, but certainly by the last half of the sixteenth century it was in general use. In conjunction with stopcocks, the common pump could supply seawater for washing decks, cleansing the hold, and fighting fires. In one form or another, it was utilized on all types of merchant ships into the nineteenth century and beyond.

Many variants of the common pump were developed, but not all were successful nor were all used on ships. The more popular maritime versions included several forms of multiple-piston pumps that appeared in the late eighteenth century. One of these was Dodgeson's pump (1799), which comprised four pistons worked by a common brake. The Deluge pump appeared in the late nineteenth century and was often found on smaller fishing and coasting vessels. The introduction of the camshaft and flywheel to pump designs in the 1840s resulted in performance standards in merchant vessels that endured into the twentieth century.

As long as the upper and lower valves of common pumps were wooden, each variety retained its own basic shape. With improvements in metallurgy during the eighteenth century, however, lead, copper, and bronze began to appear in ships' pumps, fostering different valve shapes. Iron valves were not widely used until later in the nineteenth century. Between 1810 and 1830, improved metalworking capabilities, prompted in part by the development of the steam engine, had so increased the efficiency of the common pump that it had displaced the chain pump on smaller naval vessels and on some larger warships of European and American navies.

Archaeological evidence demonstrates that the chain pump served on ships of Roman and Byzantine times. It apparently disappeared during the Middle Ages and was reintroduced from eastern Europe in the fifteenth century for land use. The idea of installing this type of pump on ships was probably brought back from the Far East by sixteenth-century European explorers. The English used it from late in the sixteenth century to the early nineteenth century, making considerable design improvements in the chain and drive wheel in the last half of the eighteenth century. By the end of this century, all European countries except France were using chain pumps on their warships. In the first decades of the nineteenth century, however, it was replaced by the common pump on some warships and by the 1850s had disappeared from them entirely. The chain pump was never used on merchant ships because of the number of men required to work it and because the maintenance of the chain made it much more labor-intensive than the common pump.

With improved technology came greater efficiency and further innovation later in the nineteenth century. Both the force pump (an old machine made better) and the new diaphragm pump appeared but were employed primarily in smaller vessels.

As was true for most shipboard chores on sailing vessels, removing the bilge water was no trivial matter. Tons of water might collect in the hold—with human muscle the only available source of pumping power. The faster and more efficiently the deed was done, especially when conditions were grave, the better it was for the ship, the cargo and passengers, and the crew. At the end of a voyage, the last task a sailing ship's crew was required to perform before being discharged was to render the hold dry. Tradition has it that the sailors could air their thoughts and feelings about the voyage, the captain, and the mates with impunity through extempore verse, as in the popular shanty "Leave Her, Johnny":

Oh, I thought I heard the old man say,
Leave her, Johnny, leave her.
You can go ashore and draw your pay.
And it's time for us to leave her.

Leave her, Johnny, leave her,
Oh, leave her, Johnny, leave her,
For the work's all done and the winds won't blow
And it's time for us to leave her.

It was maggoty beef and weevily bread,
Leave her, Johnny, leave her.
And it was "pump or drown" the old man said,
And it's time for us to leave her.

(Traditional)

Notes

INTRODUCTION
1. Unless otherwise stated, a date in parentheses after a ship's name designates the year in which the ship sank.
2. Ahlström 1978, 66–67.
3. Available from Ayers Co. Pub., Salem, New Hampshire.
4. The use of archival and patent information has been limited because of access and cost. British patents and other European naval archives and patent offices hold much promise for future research.

CHAPTER 1
1. Hutchinson 1794, 258–59.
2. Ibid.; *Treatise* 1793, 86–87.
3. *Treatise* 1793, 86–88. As a ship moves in a particular direction, greater water pressure is exerted on the foremost portion of the hull moving in that direction. Therefore, given holes of comparable size and depth in the hold, a leak is more serious in the bow than in the stern. Also, because the heel of the ship puts the leak deeper in the water, it is more severe on the lee side than on the weather side.
4. Wright 1964, 1–16; Duffy 1955, 58, 91–93; Boxer 1959, 56–58.
5. Boteler 1634, 22; *Treatise* 1793, 88.
6. Boteler 1634, 23.
7. Boxer 1959, 55–56.
8. Wright 1964, 9.
9. Boteler 1634, 23.
10. *Treatise* 1793, 89.

11. Arnold and Weddle 1978, 223, 261; Rosloff and Arnold 1984, 293, 294 fig. 8; Steffy 1994, 134, 138 fig. 5–14*b*.
12. Boteler 1634, 23.
13. Ibid.; *Treatise* 1793, 88–89.
14. Huntress 1974, 69–72.
15. Simmons 1991, 3–8.
16. Boteler 1634, 239.
17. Duffy 1955, 106.
18. García de Palacio 1587, 109.
19. Boxer 1959, 57–58.
20. Wright 1964, 10.
21. Duffy 1955, 112.
22. Ibid., 117.
23. Huntress 1974, 72.
24. Duffy 1955, 123–24.

CHAPTER 2

1. Water mains made of elm were uncovered in London in the 1930s, still in good condition after more than three hundred years. Pine tubes are known to have been used for another London water system in 1770 (Edlin 1949, 56, 126).
2. Rose 1937, 78–79; Edlin 1949, 56; O'Sullivan 1969, 107; Waddell 1984, 30–31; Ohrelius 1962, 110–11.
3. Conversation with Steven Brookes, conservator, Maine State Museum in Augusta, September 20, 1982.
4. Rose 1937, 79.
5. O'Sullivan 1969, 108; Rose 1937, 80–81; Horsley 1978, 230 fig. 96; Salaman 1975, 297 fig. 444*a*.
6. Smith 1974, 211–12.
7. O'Sullivan 1969, 107–109; Rose 1937, 80. See Horsley (1978, 225–26) for methods of marking out and shaping masts and spars. These were probably the same methods used by pump makers to give ships' tubes an octagonal cross section. Land pumps in the Mercer Museum (Doylestown, Pennsylvania) and in the Pump Museum (Saco, Maine) have square sections. The bottoms of the two pumps from *El Nuevo Constante* (1756) are the only examples of asymmetrical cross sections known to the author (Pearson 1981, 20–21 and fig. 10; Pearson and Hoffman 1995, 144–46). Their shape is most likely a product of the way in which they were seated.
8. O'Sullivan 1969, 105.
9. Salaman 1975, 39.
10. Rose 1937, 82–83.
11. Salaman 1975, 39.

12. O'Sullivan 1969, 105.
13. Ibid., 103–106, 109; Rose 1937, 81–84.
14. O'Sullivan 1969, 109.
15. Boudriot 1974, 146.
16. Salaman 1975, 296.
17. Belidor 1737–53, vol. 2: pl. 5.
18. Edlin 1949, 57, 126.
19. Waddell 1985, 250.
20. Bugler 1966, 80.
21. Dampier 1729, 443–44.
22. Telephone conversation with John Berrier, 1982.
23. Petrejus 1970, 109; Archives of the United States, Commander John Rodgers to James Strong, September, 1834.
24. Hutchinson 1794, 256–57. In this type of pump, a flexible diaphragm is placed between, but not in line with, the upper and lower valves. Hutchinson's diaphragm pump was a form of force pump with a flexible leather, the diaphragm, acting in place of the piston.
25. Björling 1890, 21.
26. Rose 1937, 87.

CHAPTER 3

1. Agricola 1556, 176.
2. Ibid.; Manwayring 1644, 78.
3. Ewbank 1842, 214.
4. Agricola 1556, 176–78.
5. Boteler 1634, 71–72; Manwayring 1644, 78–79.
6. Ladle 1993, 21.
7. Waddell 1984, 30.
8. Ibid., 31.
9. Lovegrove 1964, 117–18.
10. Carpenter, Ellis, and McKee 1974, 5 fig. 3; Redknap 1984, 29–32.
11. Oertling 1989b, fig. 3 facing 246, 249.
12. Smith et al. 1995, 26–29.
13. Watts 1993, fig. 10.
14. Waddell 1985, 257.
15. Agricola 1556, 176–77.
16. Smith 1627, 8–9; Boteler 1634, 71–72; Manwayring 1644, 78.
17. Hocker 1991, 202, 208–10.
18. Ewbank 1842, 214. During a conversation on September 9, 1982, in Ottawa, Ontario, Canada, Walter Zacharchuk, a staff member of the Canadian Parks Service, described a very simple burr pump he had seen working. The tube was a piece of PVC (polyvinyl chlo-

ride) pipe, and a trimmed mop served as the spear and upper valve. This configuration is similar in concept to that of the *San Juan*'s pump.
19. Conversation with Kevin Crisman, 1984.
20. Boteler 1634, 72; Manwayring 1644, 79.

CHAPTER 4

1. Prager and Scaglia 1972, 47–50.
2. Shapiro 1964, 566–68.
3. Akerlund 1951, 153 and pl. 16*d*.
4. Svenwall 1994, 81 figs. 51–52.
5. O'Sullivan 1969, figs. 7–12.
6. Howard 1979, 192 fig. 295.
7. O'Sullivan 1969, 111–13; Boudriot 1974, 146.
8. Ewbank 1842, 224; O'Sullivan 1969, 116; Boudriot 1974, 146; Cederlund 1981, pl. 24; ibid. 1982, 97.
9. O'Sullivan 1969, figs. 7–10, 12.
10. Telephone conversation with James Levy, conservator, Florida State Museum, 1984.
11. Cederlund 1981, pl. 24; ibid. 1982, 97.
12. John Harland to author, August 19, 1982; Cederlund 1982, 108; Bugler 1966, pl. 21.
13. Nelson 1983, 297; Cain 1983, 82.
14. Boudriot 1974, 146 figs. 184, 186.
15. Cederlund 1982, 123–24.
16. Coleman 1988; McKay and Coleman 1992, 80–81.
17. Newfoundland Marine Archaeological Society 1979, 16.
18. Oertling 1991, 14.
19. Wingood 1986, 155–56.
20. Pearson 1981, 20–21.
21. Boxer 1959, 57 and n. 1; Duffy 1955, 59.
22. Conversation with Walter Zacharchuk on September 9, 1982, in Ottawa, Ontario, Canada.
23. Wright 1964, 9.
24. Huntress 1974, 69.
25. Newfoundland Marine Archaeological Society 1979, 16.
26. Pearson 1981, 20–21; Louisiana Division of Archaeology, artifact nos. 16CM 112-1367 and 16CM 112-1186.
27. Atherton and Lester 1982, 15 and fig. 11.
28. David Beard to author, 1987.
29. Navarrete 1846, 360–65.
30. Fernández Duro 1895, 327.
31. Agricola 1556, 176.

32. Usher 1957, 326. Usher also refers to pumping applications on land, usually regarding large water systems designed to supply drinking water for part or all of a city. For the average common pump on land, the tube was probably still made of wood; other components, such as the spear and possibly the valve boxes, may have been metal.
33. Björling 1890, 7–8.
34. Oertling 1989a. The first was found by Caribbean Ventures, Inc.; the second was discovered by the Institute of Nautical Archaeology (MRW no. 598).
35. After examining photographs of one of the objects, Ivor Noël Hume and William Kelso both share the author's opinion that it is a pump valve.
36. Agricola 1556, 176.
37. Diderot 1966, 12: 787, 29: s.v. *plombiere* and pl. 2, fig. 27.
38. Ferguson 1939.
39. Ibid., fig. 5.
40. Oertling 1988.
41. Graeme Henderson, pers. comm., March 25, 1981.
42. Conversation with Dick Swete on October 5, 1982, in Gloucester, Virginia.
43. Diderot 1966, 12: 784, 29: s.v. *plombiere*, pl. 4.
44. Jones 1831, 136–37.
45. Boudriot 1974, 145, 148, and fig. 185.
46. Alan Roddie to author, June 11, 1982, with accompanying illustration.
47. Boudriot 1974, 145.
48. Howard 1979, 193 fig. 295.
49. Ibid., 193 fig. 295. Boudriot (1974, 146–48) addresses only wooden valves.
50. Henderson 1980, 239–41; Coleman 1988, 202 fig. 2*a*.
51. Two bronze valves recovered by Seaborne Ventures, Inc., of Miami, Florida, from a wreck somewhere near the island of Dominica (Voynick 1980, 69) are very similar to the valves of the USS *Hartford* and the USS *Cumberland*. Since Voynick's article was published, however, no further information regarding the date or nationality of the wreck has become available.
52. Regarding the *Pandora*, see Henderson 1980, 240 and Coleman 1988, 202 fig. 2*a*.
53. Wolf 1952, 583.
54. Usher 1954, 371.
55. Kerker 1961, 385.
56. Usher 1954, 371–72.

57. One of the dangers of relying on patents for evidence is that a poor design can be issued a patent as easily as a good one. Further, many pumps patented under a standard "pump" classification could have been adapted for maritime use.
58. In a reconstructed arrangement for the *Pandora*, a bronze quadrant brake found on the wreck works a vertical rack attached to the pump spear. This improved the efficiency of the pump and reduced wear on the piston valves by keeping the spear vertical (Coleman 1988, 203 fig. 3*a*, 204; McKay and Coleman 1992, 80–81).
59. Ewbank 1842, 226–27.
60. Ibid.; Bonnefoux and Paris 1859, 550–51; Mossel 1859, 281–84 and fig. 242.
61. College 1969, 569.
62. David Beard to author, 1987.
63. Coleman 1988, 202 fig. 2*a*; McKay and Coleman 1992, 80–81.
64. Telephone conversation with Charles Fithian, curator of archaeology, Delaware State Museums, April 26, 1993; Charles Fithian to author, February 16, 1993.
65. Björling 1890, 35, 134.
66. Ibid., 137.
67. Ewbank 1842, 227.
68. Based on examination of numerous ship's plans, this flywheel/camshaft arrangement became common by the end of the 1840s.
69. Bugler 1966, 80.
70. The loss of HMS *Royal George* in 1782 occurred while repair crews were fixing a stopcock (Barnaby 1968, 1–2). The French also used stopcocks to let seawater into ships (Boudriot 1974, 152). One was found on the wreck of *L'Impatiente* (Alan Roddie to author, June 11, 1982, with illustration).
71. Howard 1979, fig. 295.
72. Boudriot 1974, 150–52; Smith 1974, 131–32; *Prijs der Zee* 1980, 8–10.

CHAPTER 5

1. Falconer 1780, 221.
2. Ewbank 1842, 149, 157; Needham 1971, 666–67.
3. Boxer 1953, 121.
4. Needham 1971, 666.
5. Charnock 1800–1802, 2: 68.
6. For a summary of sites and sources, see Pulak and Townsend 1987, 36–38; Fitzgerald 1994, 205–206.
7. Prager and Scaglia 1972, 51.
8. Agricola 1556, 191–97.

9. Prager and Scaglia 1972, 51–52.
10. Howard 1979, 59.
11. Boteler 1634, 72; Manwayring 1644, 79. Boteler wrote "burrs," unlike Manwayring, who used "barres"; both, however, were referring to the same object.
12. Ewbank 1842, 156; Howard 1979, 59, 192 fig. 294.
13. Howard 1979, 59.
14. Blanckley 1755, 125.
15. Ibid., 126.
16. Howard 1979, 192 fig. 294.
17. *London Magazine* 1768, 499.
18. Falconer 1780, 222–23.
19. Oertling 1982, 114 fig. 1.
20. Falconer 1815, 360–61.
21. Ibid., 360.
22. Gardiner 1976, 100.
23. Falconer 1815, 360.
24. See Howard 1979 for plans of eighteenth-century warships.
25. Belidor 1737–53, vol. 2: pls. 1–4.
26. Chapelle 1949, plans 2, 4, 11, 16.
27. Ewbank 1842, 156–57.
28. Ibid., 157 and fig. 67.
29. Manwayring 1644, 79.
30. Hutchinson 1794, 254.
31. Huntress 1974, 69.
32. Bugler 1966, 79.
33. Ibid., 78–79; Longridge 1955, pls. 10, 24.
34. Longridge 1955, pl. 10; Bugler 1966, 79.
35. Parts of a chain pump may eventually be recovered from a 1703 wreck at Goodwin Sands, Kent, England, which could be the remains of one of three third-rate, seventy-gun British warships: HMS *Stirling Castle*, HMS *Northumberland*, or HMS *Restoration*.
36. Oertling 1991, 14–15.
37. Blanckley 1755, 125.
38. According to Davis (1929, 100–101), the chain loop of any given chain pump passed over two drive wheels on the gun deck and, at the bottom of the ship, through a space between the keelson and the keel. A plan of the *Raleigh* (1776), however, shows a single drive wheel on the centerline of the ship (Chapelle 1949, plan 3). Its chain may have passed under the keelson, but paired pumps, one to each side of the keelson, were the standard arrangement.
39. Conversation with John Sands on October 7, 1982, in Newport News, Virginia.

40. College 1969, 219.
41. The ships that were sunk in the York River during the siege have been the subject of extended research by the state of Virginia (e.g., Broadwater 1980, 1989; Broadwater et al. 1985, 1988). In the summer of 1980, the Nautical Archaeology Program at Texas A&M University, in conjunction with the Virginia Research Center for Archaeology, conducted an underwater excavation of the wreck designated GL-136, which was thus identified as the *Charon* (Steffy et al. 1981). The *Charon*'s chain pump has been described in detail (Oertling 1982: 116–24).
42. Steffy et al. 1981, 139.
43. Falconer 1780, pl. 8.
44. Ferguson 1939, plan facing fig. 13.
45. Longridge 1955, pl. 10.
46. McKay and Coleman 1992, 78–79.
47. Bugler 1966, 79; Longridge 1955, pl. 10; Commander John Bingeman, R.N., to author, March 23, 1983, with accompanying illustration.
48. Falconer 1780, 222.
49. Longridge 1955, pl. 10; John Bingeman to author, March 23, 1983, with illustration; conversation with Bruce Thompson, College Station, Texas, 1986; McKay and Coleman 1992, 78–79.
50. John Bingeman to author, March 23, 1983, with illustration.
51. Ibid.
52. Falconer 1815, 360.
53. McKay and Coleman 1992:78–79.

CHAPTER 6

1. Smith and Winchester catalog 1890, 99; Goulds catalog 1888, 126–29.
2. Hyde Windlass catalog 1902, 151.
3. Oleson 1984, 301–25.
4. Shapiro 1964, 567.
5. Smith and Winchester catalog 1890, 100.
6. Ibid., 101.

Bibliography

Agricola, Georg. 1556. *De re metallica.* H. C. and L. H. Hoover, trans. Reprinted 1950. New York: Dover Pub.

Ahlström, Christian. 1978. Documentary research on the Baltic: Three case studies. *The International Journal of Nautical Archaeology and Underwater Exploration* 7: 59–70.

Akerlund, Harold. 1951. *Fartygsfynden, i den farna hamnen i Kalmar.* Uppsala, Almquist & Wiksells: Boktryckeri ab.

Archives of the United States. Office of Naval Records and Library, U.S. Navy 1775–1910. Record Group 45: Fittings and equipment of U.S. ships. AF box no. 62.

Arnold, J. Barto, III, and Robert Weddle. 1978. *The nautical archaeology of Padre Island: The Spanish shipwrecks of 1554.* New York: Academic Press.

Atherton, Ken, and Shirley Lester. 1982. Sydney Cove (1797) site work 1974–1980: An overview. *The Bulletin of the Australian Institute for Maritime Archaeology* 6: 1–18.

Barnaby, Kenneth C. 1968. *Some ship disasters and their causes.* South Brunswick, N.Y.: A. S. Barnes and Co.

Belidor, Bernard F. de. 1737–53. *Architecture hydraulique, ou l'art de conduire, d'belever et de manager les eaux pour les differents besoins de la vie.* Vol. 2. Paris: C. A. Jombert.

Björling, Philip R. 1890. *Pumps: Historically, theoretically, and practically considered.* New York: E. & F. N. Spon.

Blanckley, Thomas R. 1732. A naval expositor. Mss. held in the Peabody Maritime Museum, Salem, Mass.

———. 1755. *A naval expositor, shewing and explaining the words and terms of art belonging to the parts, qualities, and proportions of building, rigging, furnishing, & fitting a ship for sea.* London: E. Owen.

Bonnefoux, Pierre M. J., and Edmund Paris. 1859. *Dictionnaire de marine à voiles et à vapeur*. Paris: Arthus Bertrand.

Boteler, N. 1634. *Boteler's dialogues*. W. G. Perrin, ed. Reprinted 1929. Publication of the Navy Records Society. Vol. 65. London: Navy Records Society.

Boudriot, Jean. 1974. *Le vaisseau de 74 canon: Traité pratique d'art naval*. Vol. 1, pt. 2. Grenoble: Éditions des Quatre Seigneurs, Collection Architecture Naval Française.

Boxer, C. R. 1953. *South China in the sixteenth century*. Hakluyt Society Pub., 2nd ser., no. 106. London: Hakluyt Society.

———. 1959. *The tragic history of the sea: 1589–1622*. Bernardo Gomes de Brito, compiler. Hakluyt Society, 2nd ser., no. 112. Cambridge, England: Cambridge University Press.

Broadwater, John D. 1980. The Yorktown shipwreck archaeological project: Results from the 1978 survey. *The International Journal of Nautical Archaeology and Underwater Exploration* 9: 227–35.

———. 1989. Merchant ships at war: The sunken British fleet at Yorktown, Virginia. In *Underwater archaeology proceedings from the Society for Historical Archaeology conference*. J. Barto Arnold III, ed. Pp. 121–23. Pleasant Hill, Calif.: The Society for Historical Archaeology.

———, Robert M. Adams, and Marcie Renner. 1985. The Yorktown shipwreck archaeological project: An interim report of the excavation of shipwreck 44YO88. *The International Journal of Nautical Archaeology and Underwater Exploration* 14: 301–14.

———, John William Morris III, and Marcie Renner. 1988. The Yorktown shipwreck archaeological project: An interim report on the 1987 season. In *Underwater archaeology proceedings from the Society for Historical Archaeology conference*. James Delgado, ed. Pp. 13–17. Pleasant Hill, Calif.: The Society for Historical Archaeology.

Bugler, Arthur R. 1966. *HMS* Victory: *Building, restoration, and repair*. Vol. 1. London: Her Majesty's Stationery Office.

Cain, Emily. 1983. *Ghost ships.* Hamilton *and* Scourge: *Historical treasures from the War of 1812*. New York: Beaufort Books.

Carpenter, Austin C., K. H. Ellis, and J. E. G. McKee. 1974. *The Cattewater wreck*. Maritime Monographs and Reports, no. 13. Greenwich: The National Maritime Museum.

Cederlund, Carl O. 1981. *Vraket vid Älvsnabben, fartygets byggnad*. Projektet Undervattensarkeologisk Dokumentationsteknik, Rapport 14. Stockholm: Statens Sjöhistoriska Museum.

———. 1982. *Vraket vid Jutholmen fartygets byggnad*. Projektet Undervattensarkeologisk Dokumentationsteknik, Rapport 16. Stockholm: Statens Sjöhistoriska Museum.

Chapelle, Howard I. 1949. *The history of the American sailing navy*. New York: Bonanza Books.

Charnock, J. 1800–1802. *An history of marine architecture*. 3 vols. London: Faulder et al.
Coleman, Ronald A. 1988. A "Taylor's" common pump from HMS *Pandora* (1791). *The International Journal of Nautical Archaeology and Underwater Exploration* 17: 201–204.
College, J. J. 1969. *Ships of the Royal Navy: An historical index*. Vol. 1. Newton Abbot, Devon: David & Charles.
Dampier, William. 1729. *A new voyage round the world*. 7th ed., corrected. London: James and John Knapton.
Davis, Charles G. 1929. *Ships of the past*. New York: Bonanza Books.
Diderot, Denis. 1966. *Encyclopédie ou dictionnaire raisonné des sciences, des arts et des métiers*. Stuttgart-Bad: Cannstatt, Frommann.
Duffy, James. 1955. *Shipwreck and empire, being an account of Portuguese maritime disasters in a century of decline*. Cambridge, Mass.: Harvard University Press.
Edlin, H. L. 1949. *Woodland crafts in Britain*. Reprinted 1973. Newton Abbot, Devon: David & Charles.
Ewbank, Thomas. 1842. *A descriptive historical account of hydraulic and other machines for raising water*. Reprinted 1972. New York: Arno Press.
Falconer, William. 1776. *An universal dictionary of the marine*. London.
———. 1780. *An universal dictionary of the marine*. Reprinted 1970. Newton Abbot, Devon: David & Charles.
———. 1815. *A new universal dictionary of the marine*. Revised by W. B. Burney. London.
Ferguson, Homer L. 1939. Salvaging Revolutionary relics from the York River. *The William and Mary College Quarterly Historical Magazine*, 2nd ser., 19: 257–71.
Fernández Duro, Cesáreo. 1895. *Armada Española: Desde la union de los reinos de Castilla y de Aragón*. Vol. 1. Reprinted 1973. Madrid: Museo Naval.
Fitzgerald, Michael A. 1994. The Ship. In *The harbours of Caesarea Maritima. Results of the Caesarea ancient harbour excavation project 1980–85*. Vol. 2: *The finds and the ship*. J. P. Oleson, ed. Pp. 205–10. BAR International Series 594. Oxford: Tempus Reparatum.
García de Palacio, Diego. 1587. *Instrucción náutica para navegar in colección de incunables americanos siglo XVI*. Vol. 8. Facsimile edition, 1944. Madrid: Ediciones Cultura Hispánica.
Gardiner, Robert. 1976. Fittings for wooden warships. *Model Shipwright* 17: 95–100.
George B. Carpenter & Co. catalog. 1903. Chicago: George B. Carpenter & Co.
The Goulds M'f'g Co.'s illustrated and descriptive catalogue and price list. 1888. New York: Goulds Manufacturing Co.
Henderson, Graeme. 1980. Finds from the wreck of HMS *Pandora*. *The International Journal of Nautical Archaeology and Underwater Exploration* 9: 237–66.

Hocker, F. M. 1991. The development of a bottom-based shipbuilding tradition in northwestern Europe and the New World. Ph.D. diss. Texas A&M University.

Horsley, John E. 1978. *Tools of the maritime trades*. Camden, Maine: International Marine Pub. Co.

Howard, Frank. 1979. *Sailing ships of war, 1400–1860*. London: Conway Maritime Press.

Huntress, Keith. 1974. *Narratives of shipwrecks and disasters, 1586–1860*. Ames: Iowa State University Press.

Hutchinson, William. 1794. *A treatise on naval architecture, founded upon philosophical and rational principles*. Reprinted 1969. London: Conway Maritime Press.

Hyde Windlass Company catalog. 1902. Bath, Maine: The Hyde Windlass Co.

Jones, Thomas P., ed. 1831. *Journal of the Franklin Institute*. Vol. 8, new ser. Philadelphia: The Franklin Institute.

Kerker, Milton. 1961. Science and the steam engine. *Technology and Culture* 2.4: 381–90.

Ladle, Lilian. 1993. The Studland Bay wreck: A Spanish shipwreck off the Dorset coast. *Poole Museum Heritage Series Number One*. Poole Museum Service, Poole Borough Council.

London Magazine. 1768. Vol. 37. London.

Longridge, Charles N. 1955. *The anatomy of Nelson's ships*. Reprinted 1977. Annapolis, Md.: Naval Institute Press.

Lovegrove, H. 1964. Remains of two old vessels found at Rye, Sussex. *The Mariner's Mirror* 50: 115–22.

McKay, John, and Ronald Coleman. 1992. *The twenty-four-gun frigate Pandora, 1779*. The Anatomy of the Ship Series. London: Conway Maritime Press.

Manwayring, Sir Henry. 1644. *The seaman's dictionary*. Facsimile, 1972. Menston, England: Scolar Press.

Mossel, G. P. J. 1859. *Het Schip*. Amsterdam.

Navarrete, Martín F. de. 1846. *Disertación sobre la historia de la náutica, y de las ciencias matemáticas que han contribuido á sus progresos entre los Españoles*. Madrid: La Real Academia de la Historia.

Needham, Joseph. 1971. *Science and civilization in China*. Vol. 4, pt. 3. Cambridge, England: Cambridge University Press.

Nelson, Daniel A. 1983. Ghost ships of the War of 1812. *National Geographic Magazine* 163.3: 288–313.

Newfoundland Marine Archaeological Society. 1979. *Newfoundland Marine Archaeological Society: Annual report, 1979*. Saint John's, Newfoundland.

Oertling, Thomas J. 1982. The chain pump: An eighteenth-century example. *The International Journal of Nautical Archaeology and Underwater Exploration* 11: 113–24.

———. 1988. Report on the pump well structure of YO-88. Report on file at the Virginia Research Center for Archaeology, Richmond, Va.

———. 1989a. A suction pump from an early sixteenth-century shipwreck. *Technology and Culture* 30.3: 584–95.

———. 1989b. The Highborn Cay wreck: The 1986 field season. *The International Journal of Nautical Archaeology and Underwater Exploration* 18: 244–53.

———. 1991. The Mombasa wreck excavation: The pumps. *INA Newsletter* 18.2: 14–15.

Ohrelius, Bengt. 1962. *Vasa, the king's ship*. Maurice Michael, trans. London: Cassell.

Oleson, John P. 1984. *Greek and Roman water-lifting devices: The history of a technology*. Toronto: University of Toronto Press.

O'Sullivan, John C. 1969. Wooden pumps. *Folk Life* 7: 101–16.

Pearson, Charles E., compiler. 1981. El Nuevo Constante: *Investigation of an eighteenth-century Spanish shipwreck off the Louisiana coast*. Louisiana Archaeological Survey and Antiquities Commission. Anthropological study no. 4. Baton Rouge: Department of Culture, Recreation, and Tourism (Louisiana).

———, and Paul E. Hoffman. 1995. *The last voyage of El Nuevo Constante: The wreck and recovery of an eighteenth-century Spanish ship off the Louisiana coast*. Baton Rouge: Louisiana State University Press.

Petrejus, E. W. 1970. *Modelling the brig of war Irene*. Hengelo, Holland: DeEsch.

Prager, Frank D., and Gustina Scaglia. 1972. *Mariano Taccola and his book "De Ingeneis."* Cambridge: Massachussetts Institute of Technology Press.

Prijs der Zee: Vondsten uit wrakken ban OostIndiëvaarders. 1980. Amsterdam: Rijksmuseum.

Pulak, Cemal, and Rhys F. Townsend. 1978. The Hellenistic shipwreck at Serçe Limanı, Turkey: Preliminary report. *American Journal of Archaeology* 91: 31–57.

Redknap, Michael. 1984. *The Cattewater wreck: The investigation of an armed vessel of the early sixteenth century*. BAR British Series 131. Oxford, England: British Archaeological Research.

Rogers, Stan. 1979. "Barret's Privateers," on the phonograph album *Fogarty's Cove*. Fogarty's Cove Music. Manufactured and distributed by Fogarty's Cove Music and Cole Harbor Music, Dundas, Ontario. U.S. distributor Silo/Alcazar, Waterbury, Vt.

Rose, Walter. 1937. *The village carpenter*. Reprinted 1952. Cambridge, England: Cambridge University Press.

Rosloff, Jay, and J. Barto Arnold III. 1984. The keel of the *San Esteban* (1554): Continued analysis. *The International Journal of Nautical Archaeology and Underwater Exploration* 13: 287–96.

Salaman, R. A. 1975. *Dictionary of tools used in the woodworking and allied trades, c. 1700–1970.* New York: Charles Scribner's Sons.

Shapiro, Sheldon. 1964. The origin of the suction pump. *Technology and Culture* 5: 566–74.

Simmons, Joe J., III. 1991. *Those vulgar tubes: External sanitary accommodations aboard European ships of the fifteenth through seventeenth centuries.* Studies in Nautical Archaeology, no. 1. College Station: Nautical Archaeology Program, Anthropology Dept., Texas A&M University.

Smith, C. F. 1974. *The frigate* Essex *papers: 1798–1799.* Salem, Mass.: Peabody Museum of Salem.

Smith, John. 1627. *A sea grammar.* Facsimile, 1968. New York: Da Capo Press.

Smith, Roger C., James Spirek, John Bratten, and Della Scott-Ireton. 1995. The Emanuel Point ship: Archaeological investigations, 1992–1995, preliminary report. Bureau of Archaeological Research. Division of Historical Resources, Florida Department of State.

Smith and Winchester illustrated catalog. 1890. Boston: Smith and Winchester.

Steffy, J. Richard. 1994. *Wooden shipbuilding and the interpretation of shipwrecks.* College Station: Texas A&M University Press.

——— et al. 1981. The *Charon* report. In *Underwater archaeology: The challenge before us. The proceedings of the twelfth conference on underwater archaeology.* Gordon P. Watts, Jr., ed. Pp. 114–43. Fathom Eight special pub. no. 2. San Marino, Calif.: Fathom Eight.

Svenwall, Nils. 1994. Ringaren. *Ett 1500-talsfartyg med arbetsnamnet.* Stockholm: Arkeologiska Institutionen.

A treatise on the theory and practice of seamanship. 1793. London: Printed for G. G. and J. Robinson, Paternoster Row, and Gilbert, Wright, and Hooke, no. 148 Lenden Hall Street.

Usher, Abbot P. 1954. *A history of mechanical inventions.* Revised edition. Cambridge, Mass.: Harvard University Press.

———. 1957. Machines and mechanisms. *A history of technology.* Vol. 3, pp. 324–46. C. Singer et al., eds. Oxford, England: Clarendon Press.

Voynick, Steve. 1980. The quest for ancient gold. *U.S. Naval Institute Proceedings* 106.12: 67–71.

Waddell, Peter J. 1984. Pump remains of the 1565 Spanish Basque whaler *San Juan.* In *Underwater archaeology: The proceedings of the thirteenth conference on underwater archaeology.* Donald H. Keith, ed. Pp. 27–32. Fathom Eight special pub. no. 5. San Marino, Calif.: Fathom Eight.

———. 1985. The pump and pump well of a sixteenth-century galleon. *The International Journal of Nautical Archaeology and Underwater Exploration* 14: 243–59.

Watts, Gordon P., Jr. 1993. The Western Ledge Reef wreck: A preliminary report on investigation of the remains of a sixteenth-century shipwreck in Bermuda. *The International Journal of Nautical Archaeology and Underwater Exploration* 22: 103–24.

Wingood, Allen J. 1986. *Sea Venture* second interim report. Part 2: The artefacts. *The International Journal of Nautical Archaeology and Underwater Exploration* 15: 149–59.

Wolf, Abraham. 1952. *A history of science, technology, and philosophy in the eighteenth century*. London: George Allen & Unwin.

Wright, Louis B. 1964. *A voyage to Virginia in 1609*. Charlottesville: University of Virginia Press.

Index

NOTE: Pages with illustrations are indicated by italics.

Admiralty of the Maze, 49
Agricola, Georg, 20, 34, 35, *36, 37*, 58
air chamber, 75
alder, 10
Alert force pump, 75, *76*
Älvsnabben (wreck), 29
anchor: capstan, 58; stock, 6
ascent tube. *See* round chamber
ash (wood), 14, 29
atmospheric pressure, 23–24
auger, 11, *12*, 13. *See also* bit
auger shaft, 13
auger stool, 13

back case, 60, 61, 63, 64, 68–69
bad air, 7
bailing, 6, 7
basal ring, 26, 44, 50, 51
Battle of the Saints, 5
beech, 10
Belidor, Bernard F. de, 13
Bentinck, John, 59. *See also* Cole-Bentinck chain pump
Bermuda, 19, 31
bilge pump, 74, 79. *See also specific pump types*

bilges, 6
bilge water. *See* living conditions, aboard a sailing ship; sanitation, aboard a sailing ship; water: bilge
biscuit, 32
bits, 11, *12*, 79
Blankley, Thomas, 59, 63
Blundell and Holmes, 51
boards. *See* planks
bolster, 63
bonnets, 58
Boteler, Nathanial, 16, 21, 58, 63
Boudriot, Jean, 29
boxes. *See* valves
brake, 16, 22, 29, 45, 74; bronze, 30, 88n. 58; for diaphragm pump, 78; double action, 46, 48; stored in pump tube, 29; variations of, 29
brass, 34, 61, 76
Brest, Fr., 41, 44
Brindley, James, 14
British Admiralty, 73. *See also* British navy
British navy: and multiple piston pump, 48, 49, 51, 52; and chain pump, 57, 58, 57, 61
bronze, 41, 45, 52, 55, 74, 80; chain

bronze *(continued)*
    links, 70; piston tube, 41–44, 45, 48; rod, 48; valve, 37, *43*, 44, 48. *See also* brake: bronze; multiple piston pump
bucket, bailing, 6, 7, 9
Bugler, Arthur, 53
burr, 16, 79
burr pump, 16–21, 79
burr valve, 16, *17*, 18, *20*, 21; of chain pump, 58, 59

camshaft, 51, 52, 80
cannon, 8, 34, 45
cannonball, 29
canvas, 5, 30
cargo, 8, 31, 81
carpenter, 5, 6, 15, 60
casks, 5
cast body. See *corps de fonte*
Castine, Maine, 10
Cattewater wreck, 18
Centaur, HMS, 5, 32
chain pump, 56–73, 80; chain of, 56, 58–61, 63, 66, 69–72; damaged, 6, 33; efficiency of, 59, 61, 62, 80; maintenance of, 58, 60, 62, 81; oldest example of, 65; seating of, 64, *66*, 67–69; used by the ancients, 58, 80; used by the Chinese, 56–58; used in Europe, 41, 45, 56, 58, 59
Challenge force pump, 75, 76
channels, on heel of tube, 30. *See also* groove
Charon, HMS, 30, 63, 66–70, 71, 73
cheeks, 22, 29
cistern: on chain pump, 56, 61, 69; on wash pump, 53, 54
claque: on burr pump, 16, 17, *18*; on common pump, 23, 24, 26, 35; on diaphragm pump, 78; keeper, 50–51; weight, 25, 27, 51

clog, 30, 31–33, 51
coal, 4, 32
cog, 47. *See also* tooth
Cole, William, 59
Cole-Bentinck chain pump, 57, 59–61, 66–70, 73
Collins, William, 61
common pump, 22–55; appearance of, 23, 54, 79; efficiency of, 20, 25, 45, 56, 80; secondary uses of, 52–54
Continental Navy, 61
copper: piston cylinder, 40, *41*, 44, 45; pump, invented by Spanish, 34; sheet, 60, 69; sieve, 30; use of, in pumps, 34, 41, 45, 52, 55, 61, 74, 80
Cornwallis, General Lord, 37, 66
Cornwallis fleet, 66
corps de fonte, 41–44
cotter key, 59, 60, 69. *See also* key
counterweight, 60
cowhide, 14. *See also* leather
crank. *See* winch
crew, 3, 4, 81
critical distance, 23
Cumberland, USS, 45
Cyane, USS, 61
cylinder: diaphragm, 75; upper valve, 37, 39, 40, 44, 45, 51

da Cruz, Gaspar, 56
dale, 30, 35, *38*, *39*
Dampier, William, 13
Daunton, Jonathan, 51
da Vinci, Leonardo, 13
De Braak, HMS, 33, 45, 49–51
debris, 5–6, 30
Defence (privateer), 10, 24, 29, 30
Delaware, 49
Deluge pump, 74, 75, 80
de Luna, Tristan, 19
descent tube. *See* back case
diaphragm pump, 14, 74, 75, 77, 78, 81

disk: of leather, 18; of metal, *34*, 35, 59, 60, 72
Dodgeson's pump, 46, 80
dogs, 10
double piston pump. *See* multiple piston pump
drive wheel, 38, 59, 61, 62, 70, 80, 89n. 38
drum, 48. *See also* roller

Edson diaphragm pump, 77, 78
elm, 10, 13, 14, 25, 29
El Nuevo Constante, 31, *32*, 33, 84n. 7
Emanuel Point wreck, 19
England, 13
esses. *See* S-link
Essex, USS, 10
European navies, and use of chain pump, 72. *See also specific navies*
Ewbank, Thomas, 51, 52, 61

facet, 30
Falconer, William, 22, 56, 70, 72
Fernández Duro, Cesáreo, 34
fid, 63
fife rail, 52
fire fighting, 54, 55, 75
flax, 26
Fleet of 1554, 5
floor timber. *See* frame
flywheel, 52, 80
foot valve, 14, 16, 17, *18*, *19*, 21, 79
force pump, 23, 85n. 24; ancient, 23, 74; for fire fighting, 54; in nineteenth century, 74, 76, 81; in Scandinavia, 23; from wrecks, 23
Fort Louisbourg, Nova Scotia, wreck at, *42*, *43*, 44
44–YO–88, 39
Fowey, HMS, 63, 66
frame, 18, 30, 53, 68
France, 80
French navy, 72

García de Palacio, Diego, 7
gasket, *28*, 30, 50, 52. *See also* leather; rubber
General Butler (canal schooner), 21
Germany, 23
Gonzales de Mendoza, Juan, 57
gouge, 30
Goulds Manufacturing Company, 74
groove, 26, *32*, 33
guidelines, 10
guns. *See* cannon

hammocks, 8
Hampton Roads, Va., 40
Hartford, USS, 45
hemp, 4
Highborn Cay, Bahamas, wreck at, 18
hole: in hull, 4, 5; pilot, 11, 79; for pump seating, 18, 19, 30, 63; in valve, 25, 26, 35, 69
Holland, 49
hooked pole, 12, 25, 60
hoops. *See* straps
Howard, Frank, 54, 59
Hutchinson, William, 3, 14
Hyde Windlass Company, 74

idler drum. *See* roller
indentation. *See* pump seating
Independence, USS, 61
India, 32
industrial manufacture, 74
intake valve, 75
iron, 34, 52, 55, 74; cast, 59, 60, 61, 67, 70; in chain pumps, 59, 63, 65, 69; in common pumps, 22, 35, 80; plate, 64; strap, 11, 29, 33, 41. *See also* brake; spear
Isle La Motte, Lake Champlain, wreck at, 21
Italy, 54, 58, 72, 79

Jamestown, Va., 31

Index

Jutholmen wreck, 29

Kalmar Find V (wreck), 23
keel, 20
keelson, 18, 19, 20, 30, 68, 89n. 38
key, 35. *See also* cotter key

larch, 10
lathe, 25, 50
lead: dale, 30, 37; lining cistern, 53; to plug leaks, 5; in pumps, 35, 37, *38, 39*, 40, *41*, 44, 45; in sieves, 30, 31, *32*, 33, 55; in tubes, 35, 39, 40; in valves, 34, 35
leaks, 3–5, 53, 56–58, 79
leather: and burr pump, 16, 17, 18, 21; and chain pump, 33, 63, 69, 70, 72; and common pump, 26, *28*, 30, 35, 50, 52; and diaphragm pump, 85n. 24; and maintenance, 21, 25, 33, 62, 70; and paternoster pump, 58; used by ancients, 58; used to plug leaks, 5
Lelystad Beurtschip (wreck), 21
lifeboat, 8
L'Impatiente (French warship), 44, 88n. 70
lines, scribed, 25, 30
living conditions, aboard a sailing ship, 6–7
logs, 10–13, 15. *See also* tube, wooden
London, 13, 51, 84n. 1
Louds diaphragm pump, 78
lower box. *See* lower valve
lower piston rod. *See* multiple piston pump
lower piston valve. *See* multiple piston pump
lower valve, 22, 25–26, *28*, 44, 50

Machault (French warship): brake from, 29; clogged pump on, 32; sieve from, 31; tube from, 13, 30, 32; valves from, 24, 26, *28*, 29

machinery, 13, 34, 45
Manwayring, Sir Henry, 16, 21, 58, 63
Mariners' Museum, 37
marline spike, 63
mass production, 60. *See also* industrial manufacture
masts, 8, 84n. 7
mast step: from the *Charon*, 66, 67; and pump seating, 18, 19, 20
Mercer Museum, 13, 24, 29
merchants, 8
merchant ships, 3, 14, 45; and arrangement of pumps, 52, 80; and chain pumps, 62, 80
metal, use of in pumps, 34, 55. *See also* specific metals
Middle Ages, 72, 80
mines, 23, 58
Molasses Reef wreck, 34–35, 55
Mombasa, Kenya, 30, 63
multiple piston pump, 47–52, 55, 80

nails, 5, 17, 27, 29, 30
National Park Service, 37
naval service, 14
Needles, The, Isle of Wight, 70
nobles, 8
Northumberland, HMS, 89n. 35
nozzle, 40

oakum, 4, 5
Ohio, USS, 61
ordnance. *See* cannon
O'Sullivan, John, 29
outflow valve, 25

Padre Island, Tx., 5
Pandora, HMS: chain pump from, 63, 69, 70; common pump from, 30, 45, 50, 88n. 58
patents, 46, 78, 83n. 4
paternoster pump, 58
Penobscot Marine Museum, 75

Pensacola, Fl., 19
pine, 10, 13, 27
pins, 59, 60, 69
piston. *See* copper: piston cylinder; multiple piston pump; valve, piston
planks: and back case, 63, 64, 69; in ceiling, 30, 64; and chain pump seating, 67, 69; and dale, 30; in hull, 20, 53; and round chamber, 69; and seams between, 4, 5; and tubes, making of, 14, 15
plate: circular, 69; iron, 64
platform, 10
plugs, 5, 14, 53
plumb bob, 11
Plymouth, Eng., 18
Point Cloates, Austral., 40
Pomone, HMS, 63, 69, 70–72, 73
Port Royal, Hon., wreck at, 14
pump: head of, 16, 17, *19*, *31*; invention of, 34; makers of, 10, 14, 24, 29; priming of, 16, 23; seating of, 18, 19, 37, 67–69; for ships, compared to ones made for land use, 10, 13, 14, 24, 30. *See* rod; spear; well; working the pumps; *and specific pump types*
Pump Museum, 24, 29, 84n. 7
puncheons, 5

quadrant, 30, 88n. 58
rack, 30, 47
Raleigh, Sir Walter, 58
Raleigh, USS, 89n. 38
Rapid, *41*, 44, 45
Red Bay, Lab., 13
Renaissance, 16
reservoir box, 35, 37, *38*, *39*, 44
Restigouche River, Can., 13
Restoration, HMS, 89n. 35
rhoding, 60
rib, 56. *See also* frame
Ringaren wreck, 23
Rivero, Diego, 34

rivet, 70
Rochefort, Fr., 41, 44
rod, 33; common pump, 13, 16, 35, 52, 79; force pump, 75. *See also* multiple piston pump; spear
roller, 56, 64, *65*, 67
rope, 4, 58
rosary pump, 58
round chamber, 56; from *Charon*, 67, 68, 69; from *Independence*, 61; from *Santo Antonio de Tanna*, 64; from *Victory*, 63
Royal George, HMS, 88n. 70
royal pump, 41–44
rubber, 52, 75
rudder, 9
Rye Sussex, Eng., wreck at, 18

sails, 5, 9
Saint Thomas, U.S. Virgin Islands, 39
Saint Thomas pumps (RS 6 and 7), 39, 40, 44–45
sanitation, aboard a sailing ship, 6–7
San José y las Ánimas (1733 Spanish Plate Fleet), 24, 29, 35, 37, *38*, 44
San Juan (Basque whaler), 13, 17–18, 19
Santiago, 9
Santo António de Tanna (Portuguese frigate): chain pump from, *62*, 63–65, 69, 72; common pump from, 30
Saõ Thomé, 4, 32
schooner, 21, 74
Scourge, USS, 29
Seaford, HMS, 59
seal, 5, 23, 26
seamen, 6, 8
seams, 4–5, 14, 35, 40
Sea Venture, 31, 32
seawater, 52–54, 80
ships, 8, 72, 79; and sinkings, accounts of, 3, 4, 5–6, 9. *See also* merchant ships; *and specific names of ships*

shot locker, 66, 67
sieve, 30–32, 33
sinkings. *See* ships: and sinkings, accounts of
slaves, 8–9
S-link, 59, 63
sloop, *20*, 21, 61
Smith and Winchester Company, 74
social order, 7, 8
soldiers, 8
Spanish Plate Fleet of 1733, 29
spare parts, 13, 29, 31, 59, 63
spars, 84n. 7
spear: in burr pump, 16, 17, *18*, 21; in common pump, 22, 27, 35, 45, 88n. 58
sprocket, 59, *62*, 63, 72
standard (pump part), 78
staple, 22, 25, 26. *See also* stirrup
stave, 5, 6, 14
steam engine, 45, 80
Stirling Castle, HMS, 89n. 35
stirrup, 44, 48, *49*, 63, *64*. *See also* staple
Stone and Depthford pump, 51
stopcock, 53, 54–55, 80
stores, 31
straps, 11, 14, 33
string, 10–12, 33
Studland Bay wreck, 17
stun'sails, 58
suction pump, 21, 22–55, 39, 74; seating of, *31*, 64, 66
Sydney Cove wreck, 33

Taccola, Mariano Jacopo, 22, 58
tacks, 5, 26, 29, 30
tallow, 4, 60
tar, 60
Tartars, 58
Taylor's pump, 47
technology, 34, 79
Texas A&M University, Nautical Archaeology Program at, 90n. 41

Thrum Shoals, Halifax, Nova Scotia, 49
tin, 32
Titus Burroughs, 40
tongs, 60
tool box, 63
tools, 14, 15, 60
tooth, 30. *See also* cog
Toulon, Fr., 41, 44
Tribune, HMS, 48–49
Trinity Cove wreck, 30, 33
tubes: boring of, 13, 45; of bronze, 41–44, 45, 48; of copper, 40, 44, 45, 61; diameter (of bore), 45, 74; heel of, 14, 30; of iron, 45; of lead, 34, 35, 37, 39, 40; PVC, 85–86n. 18; spares, 13; wooden, for burr pump, 13; wooden, for chain pump, 64, 67–69; wooden, for common pump, 10, 13–14, 29, 30–33, 41; wooden, maintenance of, 21; wooden, manufacture of, 10–15, 23, 79, 84n. 7. *See also* cylinder; planks
Turks & Caicos Islands, 34

upper valve, 27, *28*, *35*; of burr pump, 19, 22, 23, 26; of common pump, 26–27, 44, 55; spear attachment of, 27
U.S. Navy, 14, 61

valve, 13, 14, 21, 22, 24, 29; bronze, 37, *38*, *43*, 44, 45; of chain pump, 56, 59, 60, 62–63, *64*, 69–73; gravity, 37, *38*; guide, 44; lead, 34–35; maintenance of, 21, 30, 63; of paternoster pump, 58; piston, 41, 47, 50, 88n. 58; wood, 24–29, 45, 55, 80. *See also specific valve types*
varnish, 60
Vasa (Swedish warship), 29
vessel, 4, 20, 74, 80. *See also* ship
Victory, HMS: chain pump from, 62,

69, 70, 73; common pump from, 29, 53
Virginia Research Center for Archaeology, 90n. 41

Wadell, Peter, 19
warships: chain pump of, 61–62, 73, 80, 89n. 35; common pump of, 45, 52–54. *See also specific names of ships*
Washington, George, 37
wash pumps, 53–55, 75
water: bilge, 6, 30, 53; depth in hold, 3–9, 33; for drinking, 14; for fire fighting, 52–54, 55, 80; pressure, 54, 75, 83n. 3; priming the pump with, 16, 23; and rate of discharge, 34, 56, 59, 60, 78; for washing, 52–54, 80. *See also* leaks; seawater
waterline, 4, 5
water wheel, 13, 56
well, 22; caution taken in entering, 7; debris in, 6, 32, 33; of *Charon*, 66–67; of *Santo António de Tanna*, 64; and wash pumps, 53; of 44-YO-88, 39

Western Ledge wreck, 19
wheel, 48, 56, 59, 60, 66. *See also* drive wheel; water wheel
Wilkinson, John, 45
winch, 51, 56, 59, 62
wool, 26
working the pumps, 4, 7, 37, *44*, 56, 59; chain type, 56–57, 61–62; common type, 7, 16–17, 23–24, 29, 47–48; at end of voyage, 81; with shifting brake handle, 74
wrecks, 6, 18, 31, 79, 87n. 51. *See also specific names and sites*

yardarm, 6
yarn, 26
yoke, 16. *See also* cheek
York River, Va., 37, 39, 63, 66, 90n. 41
Yorktown, Va., 37, 39, 66; wreck at, 65
Yorktown pumps: RS 2 and 3, 37, *39*; RS 6 and 7, 39, *40*, 44

Zacharchuck, Walter, 85n. 18